A Natural History of Lighthouses

Published by
Whittles Publishing Ltd.,
Dunbeath,
Caithness, KW6 6EG,
Scotland, UK
www.whittlespublishing.com

© 2015 John A. Love

ISBN 978-184995-154-8

(all colour illustrations are by the author unles otherwise stated)

By the same author:

The Return of the Sea Eagle (1983)
Eagles (1989)
Sea Otters (1990)
Penguins (1994)
A Salmon for the Schoolhouse (1994)
Penguins (1997)
Rum: a landscape without figures (2001)
A Natural History of St Kilda (2009)
The Island Lighthouses of Scotland (2011)
A Saga of Sea Eagles (2013)

Printed by MELITA PRESS, MALTA

A NATURAL HISTORY OF LIGHTHOUSES

Whenever I smell salt water, I know that I am not far from one of the works of my ancestors.

Robert Louis Stevenson, 1880

… seldom in those dismal nights do our thoughts extend to the solitary outposts of our land, where, confined to the narrow cabin of a lightship, or watching in towers perched on bleak headlands or sunken rocks, the true guardians of this country's naval greatness keep their quiet and unostentatious vigil, unthought of, because remote and unknown. What, indeed, would our country be without its lighthouses? – A rugged and inhospitable land truly.

David Stevenson, 1864

John A Love

Whittles Publishing

Barra Head, the southernmost tip of the Outer Hebrides

Contents

Preface

I have been interested in natural history since I was a small boy. Born and brought up in Inverness, the capital of the Highlands, it was only a matter of time before islands appeared on my horizon. As a family we used to picnic on the Black Isle, just 'across the ferry', but then it is not a real island, merely a peninsula of Easter Ross. I think my first island would have been the Bass Rock, a boat trip on a family holiday to North Berwick. This is now one of the largest North Atlantic gannet colonies in Britain if not the world. Back in the sixties it was only a fraction of the size – nonetheless impressive. It was there too I saw my first lighthouse.

Over the decades I have visited countless islands, and much of the coast, in search of wildlife, seabirds and sea mammals in particular. And so my tally of lighthouses has grown – as has my portfolio of lighthouse photographs. I have spent all of my life in Scotland, even living on islands now for half of it. Since my retirement I have been guiding on small expedition cruise ships, mainly round Britain, which has afforded great opportunities to add more remote lighthouses, view them from new angles and in all weathers. I have seen nearly all the Scottish lights – from land, sea and/or air – quite a few others around the rest of Britain, even some as far afield as Norway, British Columbia and New Zealand.

It is perhaps not surprising that where most lighthouses came to be built, there was often really good wildlife. But then lightkeepers have been aware of this since the vocation was first created. In their isolation keepers turned to nature not just as a handy source of food, but as a useful distraction, to add interest and intellectual stimulation to their lonely lives. Those that chose to write down their encounters have sown a rich field of knowledge and observation, which sadly these days is being increasingly ignored and forgotten.

Now that all British lighthouses – and most others abroad – have become automatic, there are no resident keepers to record the passing of the seasons, the seasonal passage of birds and other animals, even to witness first hand the full fury of winter weather. Many books have been written about lighthouses and their history – and Scotland in particular has a rich heritage in that respect. It seems appropriate that someone should attempt now to honour the contribution that lighthouses and their keepers, and even some intrepid visiting naturalists, have made to natural history and especially to bird migration. In this book I try to convey as much of this as possible in the words of these men (and they are mostly men) who experienced wildlife in a manner that few others could and – since automation – few others now ever will.

one

A Fatal Force of Nature

We steered south east by east half speed until 4.00 am of the 16th inst. Presumably passing within five miles of the north west of the Flannan Isle. We should have sighted the light at midnight on the 15th inst. Weather clear but strong, south-westerly wind with very heavy sea ... notwithstanding a good look out saw nothing of the light in the islands, which we were trying to make landfall after a very heavy passage with few sights across the Atlantic Philadelphia to Leith.

Thomas Holman, master SS *Archtor*, 29 December 1900

British waters are amongst the most menacing in the world, with a long history of shipwrecks. Without any doubt lighthouses, over centuries, have saved – and will continue to save – countless human lives. But in order to do so – and almost by definition – many lighthouses came to be erected in dangerous situations, exposed to anything that nature throws at them. And thus did those brave men, who had to keep the reassuring guiding beams burning, often find themselves coping with hazardous, even life-threatening, conditions.

Scotland's first permanent lighthouse – which still stands, though long out of use – was established in 1636 on the Isle of May, now a National Nature Reserve, in the Firth of Forth. Scotland's first paid lighthouse keeper tended the coal-fired beacon on its roof and lived below it with his family. It was still blazing nightly 150 years later when a tragedy befell the family living there; perhaps Scotland's first lighthouse disaster. One night in 1791 fumes and smoke suffocated the keeper, his wife and four of their children. Only the baby survived, sucking at her dead mother's breast – a heart-rending event to which we shall return later in the book. The Isle of May beacon was maintained for 25 years more however, until replaced by candles or oil lamps. Robert Stevenson, grandfather of the famous novelist Robert Louis Stevenson, designed the imposing new station in 1816.

But perhaps the most notorious calamity to befall any lighthouse was the mysterious disappearance of three keepers at the remote and exposed Flannan Isles in 1900 – only a year after its light had first been exhibited. The exact details may never be known, but it seems that James Ducat, Thomas Marshall and Donald MacArthur were swept away in an almighty storm. Not surprisingly perhaps, the Flannans' lighthouse became a less than popular posting and was one of the early stations to go automatic in 1971. Automation may nowadays have reduced the risks for lightkeepers but for the lighthouses themselves the forces of nature remain.

Towards the end of the 19th century, congestion and the risk of collision in the English Channel were becoming critical. More and more ships were opting for a northabout route to and from the North Sea. Robert Stevenson had died in 1850, Alan, his eldest son, died in 1865, David in 1886, and the youngest Thomas – Robert Louis' father – a year later. All had been lighthouse engineers for the Northern Lighthouse Board (NLB) so David Alan (David's son, hereafter referred to as David A) was the next obvious choice as NLB Engineer. With his younger brother Charles, the family firm were commissioned to place a lighthouse on the Flannan Isles, a place the naturalist William Eagle Clarke considered 'no wilder spot to be found in the British Isles.'

The Flannans lie 17 miles (27 km) west of Gallan Head in Lewis –and second only to St Kilda as the westernmost extremity of the British Isles. They lie close to the northabout route taken by ships heading for or leaving from the North Sea ports. The islands are also known as 'The Seven Hunters' – three outliers and four closer together to the east. It was the

View of the Flannan Isles from the sea – Eilean Mor (left) and Eilean Tighe (right)

highest point of the largest and most northerly of these – Eilean Mor (17.5 ha) – that was selected for the lighthouse.

As long ago as 1853 the Commissioners of Northern Lights had recommended to the Board of Trade that a lighthouse be built on the Flannans. Trinity House finally agreed – after an appeal – in 1892. David A Stevenson got ashore the following year and concluded how 'landing of materials will obviously be attended with considerable difficulty, for there is little protection from the Atlantic swell which is seldom at rest.'

Even in calm weather, landings proved difficult, as the island is surrounded by near perpendicular cliffs some 60 m (200 ft) high. Two landing platforms – east and west – were constructed, with giant flights of steps up the cliffs of hard Lewisian gneiss. All equipment then had to be got up to the site so a horse was slung ashore by crane, to haul everything along a concrete tramway to the summit. The points where the rails from east and west landings converged into one became known as 'Charing Cross'. Sadly when Billy the horse came to be taken off, he struggled out of his sling and fell to his death into the sea below.

Old photos of the work force and the horse being slung ashore

The man holding Billy in the photograph is his owner, Ian Murdo Macleod from Breasaclete on Lewis (who was to go down with his ship in the First World War). The two men to the right in front are MacArthurs – also from Breasaclete – in the middle is Niall, ex-Navy who would survive the war, while the name of the other, far right, is not known. I wonder if he is Donald, the lighthouse occasional who soon after was to be lost on the Flannans. Having been in the Royal Engineers he would have been a likely candidate for employment in the construction of the lighthouse, along with quite a few of his neighbours in Breasaclete. Donald MacArthur, a staunch Free Presbyterian, was later employed locally in building a new church.

Only three months' work was achievable at Eilean Mor in the first year due to weather conditions. At the end of the third season the Clerk of Works suddenly passed away on site. In all, the Flannan lighthouse complex took four years to complete. Only four days after the engineers and workmen had departed, and with the lightkeepers now in post, a crane and a box of ropes at the west landing stages were washed away in a furious gale – an ominous foretaste of a tragedy that was to befall the keepers only a year later. Thomas Marshall had been deemed negligent and, through his superior James Ducat, the keepers received a letter of reprimand and, it is said, a fine of five shillings. In retrospect of course such a response to this act of nature seems totally unreasonable and it must surely have been at the back of the keepers' minds on the day of the tragedy.

Nonetheless, soon afterwards, the men's employers had been happy enough to announce:

> The Commissioners of Northern Lighthouses hereby give Notice, that on the night of Thursday the 7th day of December next [1899], and every evening thereafter from the going away of daylight in the evening till the return of daylight in the morning, a Light will be exhibited from a Lighthouse which has been erected on Eilean Mor, one of the Flannan Isles. The Light will be a Group Flashing White Light showing 2 flashes in quick succession every half minute. The power of the Light will be equal to about 140,000 standard candles. The Light will be visible all round and will be elevated 330 feet above high water spring tides, and allowing fifteen feet for the height of the eye will be seen at about 24 nautical miles in clear weather. … The top of the Lantern is about 75 feet above the island.

The focus of lighthouse life was the big living-room/kitchen, with its polished coal range, a few windows and varnished walls. From this led off the tiny bedrooms, two bunks in each, the wireless-room, storerooms, workshop, engine-house and a passage to the bottom of the tower. Nearly four decades later the naturalist Robert Atkinson described the scene:

> … inside the tower one might have been anywhere; wind and rain and Atlantic surge were remote. The lantern house in brilliant hurtful glare and the echoing stone stairs had an empty laboratory-hospital cleanliness. Silence was broken by mathematically regular clangs from a bell as the driving weights slowly unwound their way down the tower and revolved the great cage of prisms up above; then, every half-hour of the night, a ratcheting noise as the weights were wound up again. The paraffin lamp burnt with a steady purr.

There were four lightkeepers, three on a two-month shift while the fourth was enjoying a month off at home on Lewis. Although Stromness in Orkney was at first favoured, the site for the shore station and keepers' quarters came to be built at Breasaclete in Loch Roag, Lewis. This also provided a sheltered anchorage for the lighthouse tender: as weather allowed, it maintained a fortnightly resupply.

In his wonderful book *Island Going* (already quoted above), Atkinson goes on to give an account of one of *Pole Star's* regular visits in July 1937:

> The red flag meaning 'East Landing' was flying from the flagstaff, beside the Northern Lights ensign with the figure of a white lighthouse on it; the Pole Star lay off to the east; Relief Day. The engine-room had steam up: the drum unwound, the bogie went clattering downhill, forked left at Charing Cross, and disappeared round the cliff slope. This was oil-landing day and a whole year's supply for the lantern – twenty barrels – was coming ashore. *Pole Star's* motor-boat came off and lay plumbed below the derrick; a lot of men wound the handles; the weighty pendulum of a pair of barrels swung and descended into the bogie. There was much shouting up and down the uni-directional telephone between lighthouse and landing. The drum up at the lighthouse began to turn, the cable tautened, the bogie moved up the cliff across the mayweed-covered wall, set above and below with watching puffins; rounded the cliff shoulder, passed Charing Cross, drew over the golf course, and disappeared within the lighthouse. The barrels were pumped out into storage tanks and the bogie came clanking back on slack cable at alarming speed … By midday it was over: the new man had come off and Charlie went away with the boatload; *Pole Star* diminished trailing thick black smoke. She would be back again before the next Relief to land coal and water.

Transfer of men and equipment, even in summer, was rarely achieved as easily as Atkinson had witnessed. Apparently during a tricky relief in 1911 the men were almost swept away with the baggage when 'unexpectedly overwhelmed' by the sea: one keeper lost hold of the mailbag and his kitbag with all his clothes. Later, he was to be reprimanded by a totally unsympathetic NLB secretary: 'The Mail Bag should invariably be kept in a place of undoubted safety until taken charge of by the keeper coming ashore.' Sitting in his warm office the man could have shown more concern for the safety of his staff, considering the reputation of the lighthouse they were trying to reach.

By 7 December 1899 the Flannans' very first keepers were already on the island getting the light operational. No doubt they soon established a routine, not dissimilar from that at any other lighthouse in the service; they would also have begun to get to know the island. The much experienced James Ducat (a 43-year-old Arbroath man) had been promoted to principal lightkeeper in 1896: but it seems, considering his young family, he had only agreed to the Flannan's posting with some reluctance. He arrived at Eilean Mor on 28 August 1899. Thomas Marshall (aged 29 and unmarried) was second assistant, fresh from his first three years in the service on the exposed rock lighthouse of Skerryvore. The third keeper who would perish was Occasional Donald MacArthur, 40 years old, from Breasaclete. He was standing in for William Ross, the first assistant, who was off sick while the fourth in the team, Joseph Moore, was on shore leave – he and Ross were indeed a lucky pair.

December 1900 was to prove a particularly stormy month and, on Saturday the 15th, a passing steamer – the *Archtor* – failed to see any light flashing. This was not reported at once, as the captain had more pressing matters to address. On 17 December, *Archtor* had been approaching her destination when she hit a rock and had to limp slowly into Leith: here the captain informed the ship's owners about the lack of a light at the Flannans. But it would be 28 December before they forwarded the information to the NLB – who had become aware of a problem two days earlier anyway. (As if blighted by the tragedy, *Archtor* herself was to disappear without trace on 2 January 1912, somewhere off the northeast American coast.)

The west landing below the lighthouse, Eilean Mor, Flannan Isles

The NLB supply ship *Hesperus* had been due to relieve the keepers on 20 December, but bad weather had prevented it sailing until Boxing Day. It was said that during the short voyage the relief keeper Joseph Moore was restless, filled with foreboding and refused any breakfast. When at noon the vessel did arrive at the Flannans it was immediately obvious to the captain, crew and to Moore that all was not well. Captain Harvie fired a rocket but still with no response. No one appeared to greet the tender as it approached the landing. With some difficulty Moore managed to jump ashore and proceeded up to the lighthouse. The second mate and a seaman had now joined Moore, but a thorough search of the station found no trace of the keepers.

Local legend even has it that, on his first opening the door, Moore saw three shags or cormorants take off from the top of the lighthouse and fly out to sea – fiction no doubt, but ominous enough for WW Gibson to enhance his melodramatic verse:

> *We saw three queer black ugly birds –*
> *Too big by far in my belief, for guillemot or shag –*
> *Like seamen sitting bolt upright*
> *Upon a half-tide reef.*
> *But as we neared they plunged from sight*
> *Without a sound or spirit of white.*

In actual fact Moore's balanced account, written on 28 December, makes no reference to any birds. NLB Superintendent Robert Muirhead had last visited Flannan on 7 December but quickly returned after the tragedy on the 29th. His detailed analysis, together with Moore's report – with other relevant documents – are readily available on the NLB website and in the Scottish National Archives.

At the lighthouse, all but the kitchen door was closed and the fire had not been made for several days. Inside the clock had stopped and the beds were unmade. But everything else was in order, the lamps had been cleaned and refilled, while the blinds had been drawn around the lantern. It is usually said that there was an uneaten meal on the table, with an over-toppled chair beside it but Moore clearly states that: 'The kitchen utensils were all very clean, which is a sign that it must be after dinner sometime they left.' Significantly there is no mention of any chair.

Moore was obviously shocked and, in any case could not maintain the light on his own. So Captain Harvey left two seamen and the Buoymaster off *Hesperus* to assist, while the ship returned to Breasclete on Loch Roag to notify NLB Head Office in Edinburgh. Harvey's telegram read:

A dreadful accident has happened at Flannans. The three keepers, Ducat, Marshall and the Occasional have disappeared from the Island. On our arrival there this afternoon there was no signs of life to be seen … signs indicated that the accident must have happened about a week ago. Poor fellows they must have been blown over the cliffs or drowned trying to secure a crane or something like that …

At the east landing Moore and his companions found everything just as it had been left at the last relief on 7 December. At the west landing however, there was chaos, dramatic evidence of the severity of the recent storms. Railings were bent, boxes and ropes strewn around and a block of stone weighing more than a ton had moved position. Two hundred feet above the sea, turf had been ripped off the cliff top.

The last entry on the slate was by Principal Lighthouse Keeper Ducat on the morning of Saturday 15 December. Moore noted that:

> Up till 13th is marked in the book and 14th marked on the slate along with part of 15th. On 14th the prevailing state of the weather was – westerly olb bry shrs [Muirhead renders this 'westerly oh by 1 hrs']. On 15th the hour of extinguishing [the light] was noted on the slate along with barometer and thermometer inside and outside lantern taken at 9am as usual and direction of wind.

The information on the slate would later have been transferred to the logbook by the principal lightkeeper but no one now knows what the logbook revealed. It is likely that Muirhead took it back to Edinburgh for the enquiry and it has now been lost.

On the other hand, a version of logbook entries reappears at a later date, which Keith McCloskey – in his well-researched *The Lighthouse: the Mystery of the Eilean Mor Lighthouse Keepers* (2014) – discusses at length. Supposedly written by Assistant Lighthouse Keeper Thomas Marshall, this recounted:

> Dec 12th: gale north by northwest. Sea lashed to fury. Never seen such a storm. Waves very high. Tearing at lighthouse. Everything shipshape. James Ducat irritable. [Later.] Storm still raging. Cannot go out. Ship passing sounding foghorn. Could see lights of cabins. Ducat quiet. McArthur crying.
>
> Dec 13th: storm continued through night. Wind shifted west by north. Ducat quiet. McArthur praying. [Later.] Noon, grey daylight. Me, Ducat and McArthur prayed.
>
> Dec 14th: [No entry.]
>
> Dec 15th: storm ended, sea calm. God is over all.

It seems that this bizarre text first appeared in a magazine *Strange True Stories* (1929) by an American pulp fiction writer Ernest Fallon – from 'English Sources'. In 1965 the text is recycled in a book entitled *Invisible Horizons: True Mysteries of the Sea* by another American Vincent Hayes Gaddis, the man who, only the year before, coined the term 'Bermuda Triangle'. Since there is no verification that Marshall's supposed entries are true, one can only conclude that they perpetuate the pulp fiction strand. Nothing suggests any basis in fact for Fallon's cryptic 'English Sources'.

It is normally the principal's duty to write the log so Marshall's comments on the behaviour of his mess-mates in an official document is immediately suspect. It is understandable that Ducat might be irritable since already his relief was several days overdue and his family

The west landing,
Eilean Mor

had been preparing to have him home for Christmas. But Ducat was usually described as being a good-natured man. It is also strange that the Occasional Macarthur was said to have been crying. It is likely he was made of sterner stuff since he had been described as quick-tempered, a bit of a brawler and deeply religious.

Only MacArthur's oilskins were found hanging indoors so Moore surmised that he had left the building in a hurry. The official conclusion was that Principal Ducat and Assistant Marshall had gone down to the landing in the storm to secure the gear and been washed away in a great wave; the third keeper then left to assist them without his oilskins.

As an example of media coverage, on 29 December 1900 the *Highland News* reported 'Another Mystery of the Deep'. The relief party:

> … found the blinds were drawn over the windows. The keepers' beds were unmade, just as they had risen from them, and their half-finished breakfast was there on the table, with the chairs pushed aside as if they had hurriedly risen and gone out. … It was discovered that one of the cranes – that on the west side – was carried away by the storm and as an oilskin was found fixed to the wreckage, it is believed the men were endeavouring to secure the crane when they were either blown or washed off.

Quite where the newspaper gleaned these details, such as blinds drawn over the windows, is perhaps the real 'mystery of the deep' since it was normal practice for the blinds in the lightroom to be drawn each morning, to prevent the sun overheating the mechanism. Moore had in fact reported that the crane was safe, and all the kitchen utensils were clean so the men must already have had their meal. One wonders why, if any of them had left in a hurry, they carefully shut the entrance gate outside, and the door into the lighthouse, while only the door into the kitchen was ajar. But maybe the wind blew them shut later. Superintendent Muirhead would soon conclude that the men could not have been blown off the cliff since the wind was westerly and would have taken them inland rather than into the sea. The *Highland News* was to correct a few of their errors in the following week's edition, by which time other Highland newspapers together with the *Scotsman* were able to be more accurate.

It was only after 1912, when an English poet Wilfrid Wilson Gibson (1878–1962) published his epic *Flannan Isle,* that the story began to assume such an air of mystery, speculation, even intrigue:

> Though three men dwelt on Flannan Isle
> To keep the lamp alight
> As we steered under the lee, we caught
> No glimmer through the night …
>
> …. Of the three men's fate we found no trace
> Of any kind in any place,

But a door ajar and an untouched meal
And an over-toppled chair …

… We seemed to stand for an endless while
Though still no word was said.
Three men alive on Flannan Isle,
Who thought on three men dead.

Thus arose the myth of the untouched meal and an over-toppled chair; but no more than deliberate poetic devices to add drama – and reminiscent of the *Marie Celeste*. Gibson even went on to mention 'how the rock had been the death of many a likely lad': but the light had only been lit for a year! Admittedly during construction the Clerk of Works had died there of natural causes, and a few years later John McLachlan would fall to his death from the balcony, but nothing else out of the ordinary seems to have happened at Flannan before or since (though it always did remain an unpopular posting amongst lightkeepers). Not surprisingly Moore had been apprehensive about returning to his post and, after ensuring the new men had settled into the routine, he was transferred elsewhere only a few weeks later. Details about him seem confused but it seems that he might have come from Fife, but originally from Belfast (and had at least one son James); he was to live to a ripe old age of 83, and died in Partick in Glasgow.

Ross's luck was short-lived. He too was transferred from the Flannans barely two months after the tragedy, first to the Sound of Mull as principal lightkeeper and then to Eilean Glas lighthouse in Scalpay, Harris where, only a year and four months after his work-mates were lost, he dropped dead in the lightroom on 15 April 1902.

The three dead keepers left two widows and six children. Marshall was unmarried. The Ducats had four of a family – Arthur (aged 6), Anna (9), Robert (13) and Louisa (16). On her 100th birthday Anna, a former nurse, recalled how her father, on departing for the Flannans, would always lift her and her little brother up to give them a kiss. But on this bright sunny day he turned and when they ran over to him he gave them a second kiss, almost as though – she surmised – he had some premonition of the impending disaster. James Ducat was then 43 and had joined the Lighthouse Service when he was 22. He would be stationed at Scurdie Ness near Montrose, Inchkeith, Rinns of Islay, Langness on the Isle of Man and Loch Ryan before he was posted to the Flannans on 28 August 1899.

On his discharge from the army seven years previously, Occasional Donald MacArthur had moved back to Breasaclete to join the Lighthouse Service two years later. He left a 32-year-old widow – originally from Gravesend – with a 7-year-old girl and 10-year-old boy. After the tragedy, Superintendent Muirhead reported to the Board that he was concerned for her future. And indeed he won for each of the families a small pension. He had known all three men:

They were selected, on my recommendation, for the lighting of such an important Station as Flannan Islands, as it is always my endeavour to secure the best men possible

Eilean Mor lighthouse and the ancient early Christian cell of St Flann

... the Board has lost two of its most efficient Keepers and a competent Occasional ... I was with the Keepers for more than a month during the summer of 1899, when everyone worked hard to secure the early lighting of the Station before winter, and, working along with them, I appreciated the manner in which they performed their work. I visited Flannan Islands when the relief was made so lately as 7th December, and have the melancholy recollection that I was the last person to shake hands with them and bid them adieu.

On that occasion Superintendent Robert Muirhead had been accompanied by his wife who may have taken the photograph – the only photo it seems – of him with the three missing keepers; it is reproduced in McCloskey's book.

Ever since, there has been wild speculation about the men's fate – most of them too fanciful to relate, from the paranormal and hauntings, sea serpents, the effects of food poisoning, hallucinogenic effects from ergot contamination in their bread causing crazed attacks on each other, to suicide, murder, kidnapping and alien abduction. The story has even featured in an early episode of *Dr Who*, while various songs and pieces of music have been written (one by the rock group Genesis, another by a New Zealand folk band Beltane); there is a full chamber opera *Lighthouse* by Sir Peter Maxwell Davis (1979), Gibson's poem of course, a beautifully illustrated children's book by Gary Crew and Jeremy Geddes, at least one novel (*A Strange Scent of Death* by Angela J Elliott) – and a video game! And so the 'mystery' is perpetuated.

But for me, and many others including lightkeepers themselves, there is no mystery and never has been. There is no need to invoke the sinister or the paranormal, it was purely a tragic act of nature. There seems no reason to doubt Superintendent Muirhead's initial assessment – the men had got swept away in the storm by abnormally rough seas. Since it was not permitted for all three to abandon the lighthouse, only two of the men – probably the pair that had been censured for previous damage to a crane and ropes – must have gone down to the landing to secure the gear. The third, Donald MacArthur, would have remained back in the lighthouse. But when his companions did not return he would have been concerned for their safety. Or else, perhaps, he saw a great wave approach and rushed to warn them, but without his waterproofs (which, in any case, he often declined to wear). MacArthur may have been too late, only then to be swept away himself.

In actual fact the immediate area of the landing is not visible from the lighthouse while dramatic spray from a blowhole might be. This may have been what fuelled MacArthur's concern. A keeper posted to the Flannans in the 1950s elaborated on this scenario, as McCloskey recounts. A former keeper thinks that initially one man was washed off the west landing and – unable to rescue him alone – the other ran all the way back to the lighthouse to fetch help from MacArthur. It does, however, seem a bit unlikely that a man in such surf would have remained within range by the time this help arrived. Another recent keeper, Jack Ross, had a particularly relevant narrow escape at the west landing. He and two companions saw a 'giant wave' come out of nowhere. They had no time to scramble up the steps to safety, only to grab any support within reach which saved their lives. Again – as McCloskey records – Ross had said the wave hit them like the proverbial brick wall and had they not been prepared for its approach there would have been another Flannans disaster to mystify the world.

The fact that the appalling incident happened only a year after the lighthouse became operational is probably highly relevant. Although the keepers had been in post for a little time before the light came to be exhibited, they may not yet have been totally familiar with winter storm conditions around the island. Having landed on Eilean Mor myself, albeit in summertime, I can easily appreciate how exposed are the cliffs. Behind the west landing stage is a great cave and perhaps it was the violent backwash out of here that caught the men unawares. Despite all the official reports and documents, theories and speculation over the years, there is no better explanation for the mystery. Keith McCloskey's recent analysis includes the historic photograph of the three keepers with Superintendent Robert Muirhead taken only a week previous to their disappearance.

No bodies were ever found. The sea is reluctant to give up its dead – or its secrets. But without doubt it can be a massive force to be reckoned with. The oceans cover just over 70% of our planet and the Atlantic is the second largest. Swell on these waters is created by wind, their resultant height determined by fetch, unhindered in the case of the North Atlantic by

Last photograph of the three keepers on Eilean Mor: from left to right, Marshall, Ducat and MacArthur with Muirhead on the right (Steven Gibbons)

3,000 miles of open sea. In addition, temperate seas are highly changeable, according to the season and climate.

Approaching the shore, rollers diminish in height and slow down due to friction. They stack up and, since the sea near the surface moves faster than against the bottom, the waves spill over as they break on the shore. With a gentle gradient offshore the breakers are less violent than against a steeply shelving shoreline such as the sheer cliffs of the Flannan Isles. Flannan keepers observed how, on stormy days, spray would hit the lantern, nearly 1,000 ft (328 m) above sea level.

What we do know is that whenever waves hit cliffs, detached boulders can be hurled at the land like missiles. Wave impact is explosive; pressurised water is forced into cracks, prising the cliff face apart, hastening erosion. The 1899 foghorn at Dunnet Head in Caithness, the northernmost point of the British mainland, has long disappeared over the eroding cliff; another had to be built, and then a third in 1952. At Muckle Flugga, the northernmost extremity of the British Isles, a perimeter wall five feet high and two feet thick was thrown down, and the door of the lighthouse broken open – at a height of 195 ft (59 m) from the sea. Similarly, in February 2013, part of the wall enclosing the South Light on Fair Isle – at sea level – was destroyed in a ferocious storm and for a time the lighthouse complex was completely surrounded by the sea. One of the occupants of the lighthouse accommodation saw the huge wave approaching and just made it to safety in time. Waves can reach 50 ft (15 m) in height, the highest recorded being 112 ft (34 m) with a speed of travel of 63 mph (102 km/h). Giant, rogue or freak waves are by no means unknown to deep-sea mariners, but what exactly causes them is still being researched.

Stormy seas, Ardvule, South Uist

One recent incident illustrates an all too common scenario, almost a replay of the Flannan disaster – but fortunately without loss of life. Not long after Dubh Artach had been automated in 1994, a technician was sent out on a routine service: 'It was notorious for its unpredictable waves. You never know from what direction they're going to come.' The helicopter finally managed to drop off the two men on the rock. They got to the foot of the ladder and turned around to see a vast wave rolling towards them. They had a mad scramble up the ladder and just got inside as the wave swept right across the rock.

Keepers maintained that pillar rock lighthouses shook under wave impact. The Bell Rock was worst of all, 'jumping' in strong northeasterlies. Even the tallest rock lighthouses can be overtopped by violent storms. Statistics for the two decades between 1881 and 1890 highlighted Skerryvore as being one of the stormiest parts of Scotland. During that period there were a total of 542 storms lasting 14,211 hours.

One Skerryvore keeper lost his hearing for several weeks, after a lightning strike had thrown him back through the entrance door, fortunately not outwards. In its first year of operation, the copper lightning conductor on Dubh Artach, at a height of 28 m (92ft) above the high water mark, was wrenched out of its sockets by a storm. Nor were lighthouses situated on high cliffs immune. Keepers at Barra Head lighthouse, 211 m (690 ft) above the sea, used to find small rocks – and even fish – thrown up and into the courtyard. In the 1850s the Gaelic folklorist Alexander Carmichael considered Barra Head to have terrific seas with no equal around the British coast. The keepers told him of several large boulders, some weighing up to 80 t, which were 'knocked about', the largest tossed 12–14 ft (c.3–4 m) from the high water mark, while a mere three-tonner was pitched some distance inland over a 60 ft (18 m) perpendicular cliff. On a visit to the same place in 1936 the geologist Archibald Geikie noticed a 43 t block of gneiss that had been moved 5 ft (1.5 m) during a storm!

Tuskar Rock lies 11 km (7 miles) off the Wexford coast in southeast Ireland, a tiny flat rock outcrop accommodating little else but the 34 m (112 ft) granite tower. Only 56 km (35 miles) from the Welsh coast, treacherous currents and tides in storms renders the Rock one of the most hazardous obstacles to shipping on the Irish coast; at least 24 wrecks are documented. When the lighthouse was being constructed in October 1812, 10 of the 24 workforce were swept off the rock as they slept and perished– all but two were married with families. The survivors were found still clinging to the wet rock two days later, but nobly went on to finish the task in hand before moving on to construct other lighthouses. A stone-cutter fell 72 ft (22 m) to his death on the same rock in August 1814.

The following year the Tuskar Rock light was finally exhibited. Its principal lightkeeper, M Wishart, would be demoted in 1821 for his involvement in smuggling. As assistant lightkeeper on the Skellig Michael he would later fall off the cliff while cutting grass for his cow. In 1956 another keeper disappeared over the cliff; whilst, still on Skellig, back in 1825 – the year before the light was exhibited – a labourer was killed blasting rock. Yet another keeper was luckier – he decided to unblock Skellig's outside privy using one of the tonite charges normally used as a fog signal; he succeeded except that the explosion took the roof off!

Few Irish lighthouses are more dangerous perhaps than the two towers built on the 5.7 ha Eagle Island off the coast of County Mayo, in northwest Ireland – a situation with many parallels to the Flannan Isles. Being only 32 km (20 miles) from the continental shelf, Eagle Island is subject to the full fury of the Atlantic. Storm waves frequently overtopped both towers, situated 68 m (223 ft) above the sea. In January 1836, a year after it was lit, a rock

Skellig Michael (left) and Little Skellig (right), County Kerry, Eire

thrown up in storm shattered the glass of the West tower – but the keepers had the light operational again within the hour. Both lanterns were badly damaged and one tower was swamped so that the keepers had to drill holes in the walls to let the water drain out before they could open the door! It took them four or five days to restore the light so that, when the weather finally allowed a technician to come out from the mainland, all he had to do was replace the glass. On 29 December 1894 the family accommodation at the East tower was damaged beyond repair, its light was extinguished and the families had to take shelter in the tower until help arrived. The following year engineer William Douglass (himself having suffered a fall off Eddystone Lighthouse only to be saved by a large wave) improved the West light so that the East Light could be extinguished once and for all. The families were moved to mainland accommodation in 1900. The West light was put out of action by a storm in 1935, again in 1987 and in 1988 when it was finally made automatic.

So the tragic fate of the Flannans' lightkeepers can be seen in context, a violent encounter with the elements, nature at its most savage and unforgiving. Incredibly, a local journalist researching the Flannan incident recently, heard a story about a woman living near the shores of Lewis, who was hanging out her washing on that fateful day, and saw a massive

Flannans lighthouse at sunset

wall of water coming in from the west. She ran back to the house just as this large wave hit the shore – but her washing and washing line were swept away!

Exactly 100 years after the Flannan tragedy, residents and community leaders in Lewis, descendants of the families and officials from the NLB, came together in Breasaclete, to hold a memorial service and hold a minute's silence in honour of the three lost lightkeepers.

two

The Greatest Storm

No pen could describe it, nor tongue express it, nor thought conceive it, unless by one in the extremity of it ... Never was such a storm of wind, such a hurricane and tempest known in the memory of man, nor the like to be found in the histories of England.

Daniel Defoe, 1804

Britain is surrounded by some of the most dangerous waters on the planet. Its coast is strewn with hazards, and shipwrecks were once commonplace. It was once said that an average of 550 ships were wrecked around the British Isles each year. One estimate in the 19th century even put the figure as high as 1,800 per year. On that basis, a ship was being driven ashore and battered to destruction about every six hours! The English Channel in particular has long been one of the most congested and perilous shipping routes in the world. The Goodwin Sands for instance, off the Kent coast, is the final resting place of over 2,000 vessels – the largest concentration of wrecks anywhere in the world – from Elizabethan galleons to U-boats.

Similarly, further west along the Channel, the Eddystone Rocks, near the entrance to the strategic port of Plymouth, were infamous. Around 1620 the master of the *Mayflower* described the obstruction as 'twenty three rust red ... ragged stones around which the sea constantly eddies, a great danger ... for if any vessel makes too far to the south ... she will be caught in the prevailing strong currents and swept to her doom on these evil rocks.' Bad weather, of course, added greatly to these perils. Although this volume is mainly about lighthouses in Scotland, it is appropriate to begin the story in the English Channel, specifically Eddystone Lighthouse, arguably the most famous lighthouse in the world. Its story is a fascinating one. The fourth incarnation (which has happily survived virtually intact and re-erected on Plymouth Hoe) pioneered successful marine engineering practices on offshore tidal reefs while its elegant, red and white form has become the archetype of everyone's idea of a seaside lighthouse.

During the 17th century Plymouth was becoming more important due to the establish- ment of a naval dockyard and to trade with the Americas. Local dignitaries pressed for a

Smeaton's Eddystone lighthouse re-erected on Plymouth Hoe

beacon on the dangerous rocks of Eddystone, 22 km (14 miles) south of Plymouth, which necessitated building on a rock with only a metre ever exposed at high tide. At the time there was no such thing as a marine engineer, but a flamboyant and eccentric entrepreneur Henry Winstanley (1644–1703) rose to the challenge. Winstanley was speculating in sea cargo and two of his own ships had come to grief on Eddystone. So he had a special reason for getting involved. His creation now seems bizarre, improbable and doomed from the start but in 1959 the lighthouse historian (and David A's nephew) David Alan (D Alan) Stevenson graciously commented:

> ... of all works connected with the sea, the erection of this tower was one of the most daring and original ever to be attempted, and Winstanley must be honoured

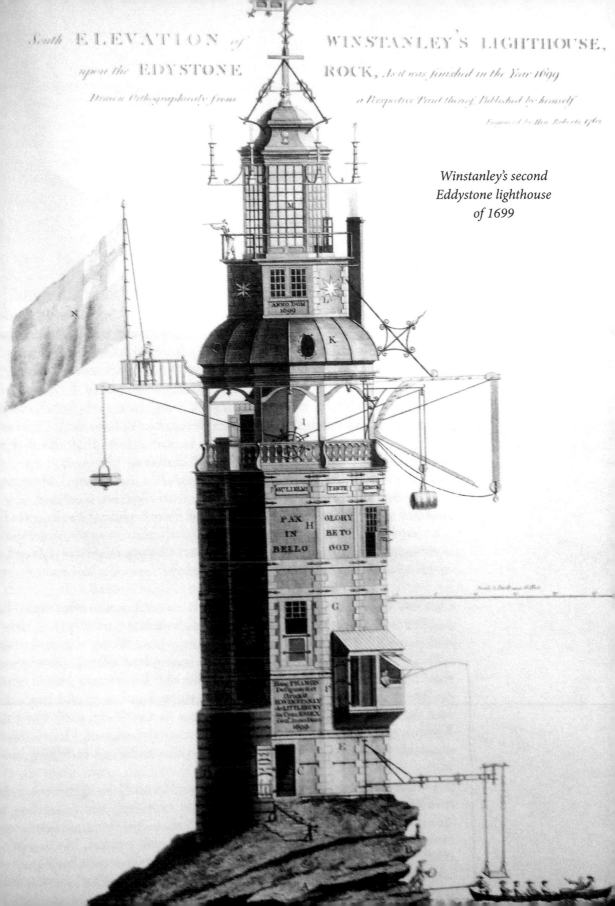

*Winstanley's second
Eddystone lighthouse
of 1699*

for his decision to build it on these wave-swept rocks, and for his ingenuity and his unflagging resolution to carry out his plan – thus faulting the universal qualified opinion as to its impracticability.

Immediately on completion, Winstanley had to modify his initial structure until in 1698 the tallow candles were finally lit atop this second, equally bizarre, tower, every bit as ornate as the first. It stood 121 ft (37 m) tall with numerous strange appendages of dubious practicality – let alone durability. Nonetheless he proudly announced that 'it is finished and will stand forever as one of the world's most artistic pieces of work'.

Indeed so confident was he that he desired to experience for himself just how his masterpiece would withstand storms and rough seas. In the early morning of Friday 26 November 1703 – still fit and strong for all his 59 years – he bravely set sail from the Barbican Steps in Plymouth with several men who were to relieve the duty lightkeepers. He himself intended to stay in the tower for a few days, to oversee some repairs before returning ashore and onwards by coach to his London home for Christmas.

However, after several weeks of persistent gales the sea was still displaying a considerable swell. The small tender had undertaken the voyage in similar conditions many times before but, due to the uncertainty of the weather, the skipper James Bound, was reluctant to take Winstanley. His esteemed passenger insisted however and, as feared, the wind freshened alarmingly just as the boat approached 'the Stone'. Inside the lighthouse, having endured weeks of bad weather, the duty keepers could not wait to depart having, at times, even feared for the safety of the structure itself. Not without risk, the relief was completed.

With the barometer plummeting and the storm worsening, Winstanley and the new shift immediately set to securing everything that might dislodge in the shaking and damaged tower, while their boat set off to battle shoreward against the elements. It was not long before the tower slowly began to disintegrate, despite the lightkeepers' desperate repairs. These proved in vain, sadly, for Winstanley's pride and joy – which he had claimed would endure forever – ultimately broke up and – with everyone trapped inside – toppled into the angry seas, never to be seen again. It is said that, back in his Essex home that fateful night, Winstanley's treasured wooden model of his Eddystone masterpiece mysteriously crashed to the floor.

It was the worst storm ever known, inflicting massive loss of life and untold damage. Much later the poet Jean Ingelow (1820–97) would recall the tragedy:

> *And the winds woke, and the storm broke,*
> *And the wrecks came plunging in;*
> *None in the town that night lay down*
> *Or sleep or rest to win.*
> *The great mad waves were rolling graves,*
> *And each flung up its dead;*
> *The seething flow was white below,*

The loss of Winstanley and his lighthouse, 1703

> *And black the sky o'erhead.*
> *And when the dawn, the dull grey dawn,*
> *Broke on the trembling town,*
> *And men looked south to the harbour mouth,*
> *The lighthouse-tower was down!*

The writer and creator of *Robinson Crusoe*, Daniel Defoe (*c.*1660–1731), assembled a ground-breaking work of journalism, published in July 1704 and imaginatively entitled *The Storm*. Defoe admitted:

'No pen could describe it, nor tongue express it, nor thought conceive it, unless by one in the extremity of it … Never was such a storm of wind, such a hurricane and tempest known in the memory of man, nor the like to be found in the histories of England.' On a new moon and high spring tides, a great hurricane had swept across the Atlantic from America with winds of 140 mph (225 km/h) being recorded in the Channel. 'Our barometer informed us that the night would very tempestuous; the mercury sank lower than I had ever observ'd it on any occasion whatsoever, which made me suppose that the tube had been handled and disturb'd by the children'.

A dozen fighting ships of the Royal Navy – returning from assisting Spain against France in the War of the Spanish Succession – were lost on the Goodwin Sands and 1,600 sailors drowned. Many other vessels suffered damage: 700 were cast adrift and crushed together in the Pool of London, while another vessel was torn from her moorings in a Cornwall harbour to end up – eight hours later and still with her crew inside – on the coast of the Isle of Wight 320 km (200 miles) away. In all at least 500 large sailing ships were overwhelmed and more than 8,000 seamen lost their lives: others have put the figure as high as 15,000.

Hundreds of people and countless more livestock were drowned in flooding on the Somerset Levels. Coastal towns such as Portsmouth looked 'a picture of desolation … as if an enemy had sackt them and were most miserably torn to pieces'. Buildings all over the south of England suffered damage and at least 123 people were killed by falling masonry or flying debris. The Bishop of Bath and Wells and his wife were crushed to death. Twenty-one others were killed by the 2,000 chimneys that tumbled in London alone. The lead roofing was torn from Westminster Abbey while Queen Anne herself had to take refuge in the cellars of St James' Palace. A hundred elms fell in St James's Park while no fewer than 4,000 oaks were toppled in the New Forest alone. As he rode round Kent in the aftermath, Defoe gave up counting at 17,000 fallen trees, writing 'I have great reason to believe that I did not observe half of the quantity.'

Defoe concluded that the country had experienced 'the greatest, the longest in duration, the wildest in extent, of all the tempests and storms that history gives any account of since the beginning of time'. To those in the thick of the storm it would certainly have felt like it, but without doubt Defoe – who was indeed there at the time – was also prone to a certain amount of what we now recognise as journalistic licence.

And no sooner had the Eddystone light been extinguished so dramatically than the brig *Winchelsea* – having crossed the Atlantic in the teeth of many gales – came to grief the very next day, with the loss of all but two of its 69 hands. In vain the captain had searched desperately through the storm for the familiar rays that would guide him into his home port of Plymouth, only for his ship to smash into the rock of Eddystone, just 22 km (14 miles) short of safety. Having come to rely upon Winstanley's beacon, mariners were once again in danger, so a replacement became a matter of urgency.

John Rudyerd (1650–1718), a silk merchant with practical bent, came forward having sought advice and co-operation from naval shipwrights. In 1708, just four-and-a-half years after the loss of Winstanley's tower, Rudyerd's builders had erected his elegant 'conical ship' of stone blocks keyed into the parent rock, stout seasoned timbers bound with iron bands, all clad in hardwood panels caulked with oakum and coated in pitch. The structure demanded considerable and costly annual maintenance but endured for nigh on half a century before its fatal weakness was exposed – fire. On the night of 1 December 1755 the man on watch, a sprightly 94-year-old (some say he was only 84) called Henry Hall, discovered that an accumulation of soot and tallow from the candles had ignited and the lantern was ablaze. Rousing his two companions he rushed back up to the gallery and began dousing the fire with buckets of water. Aghast and with mouth agape however, he apparently swallowed a

Left: Rudyerd's Eddystone lighthouse, 1708

Right: Fire at Rudyerd's Eddystone lighthouse, 1755

dollop of molten lead from the roof. No one believed him as he fought on bravely trying to dowse the fire, to no avail. Eventually the three keepers had to abandon the tower and spent the rest of the night outside on the rock until help arrived after eight hours. Hall died 12 days later and a surgeon removed a seven-ounce lump of lead from his stomach! His remarkable story was verified – the earliest proven case of lead poisoning – and the fatal 'bullet' can now be seen in the National Museum of Scotland in Edinburgh.

The man charged with yet another replacement lighthouse was a Leeds instrument maker and mathematician known to have a considerable flair for civil and mechanical engineering problems, John Smeaton (1724–92). He shied away from the acclaim his hesitant southern colleagues soon had to bestow upon him: 'His style and language had a particular and in some degree provincial way of expressing himself … a way which was very exact and expressive, though his diction was far from what might be called classical or elegant.'

With characteristic forethought Smeaton quickly realised that the shape best suited to withstanding the mighty forces of the sea resembled the flared trunk of an oak tree. The material needed was nothing less than pure stone, an interlocking structure of variously hewn blocks assembled like a three-dimensional jigsaw, further bound together by 1,800 oak pins or trenails and held with a quick-setting marine cement of his own devising. The building – at 72 ft (22 m), similar in height to Rudyerd's – had been completed by 1759 without loss of life, using 1,493 blocks of Portland stone and granite, some weighing up to two tons each (together amounting to 988 tons in total). The total cost with its lantern and 24 candle light, was around £40,000. In 1810 Trinity House would install oil lamps and 24 reflectors, which greatly extended the initial range of only five miles or so. Smeaton had finally tamed the Eddystone Rock and many of his innovative methods were to remain standard in lighthouse construction across various parts of the world for 250 years.

In the end it was not his famous tower that failed but the eroding rock upon which it stood. After a visit in 1818 the Scots engineer Robert Stevenson observed how:

> The house seems to be in very good state of repair … As to the rock … it is shaken all through, and dips at a considerable angle … And being undermined for several feet it has rather an alarming appearance … Were I connected with the charge of this highly important building … I should not feel very easy in my mind for its safety.

It was not until 1879 however, that work began on a replacement, to be designed by Sir James Nicholas Douglass (1826–98), engineer-in-chief to Trinity House. He found the lighthouse '… in a fair state of efficiency; but, unfortunately, the portion of the gneiss rock on which it is founded has been seriously shaken by the incessant heavy sea strokes on the tower, and the rock is considerably undermined at its base …'.

Douglass's task demanded building on a neighbouring rock, which was covered at all states of the tide. Back in 1811, Robert Stevenson had overcome such problems on Bell Rock (see later), and was supplied by slow sailing vessels, which was all that was available at the time. By now Douglass was able to benefit from steam to deliver the pre-hewn stone blocks (4,668 t of them), and could employ more sophisticated mechanical cranes, pneumatic tools and, last but not least, the accumulated experience of his father Nicholas – already an eminent lighthouse engineer – and his brother William. James's own son William Tregarthen was also to become a lighthouse engineer but he almost lost his life at Eddystone. He fell 21 m (70 ft) from the top of the tower during its construction. Miraculously a large wave swept in and caught him, so that he could be rescued unharmed from the sea. Not long afterwards he was to die in a boating accident.

The fifth Eddystone tower employed many structural features pioneered by Smeaton and by Stevenson on Bell Rock. It proved just as functional and elegant, yet was larger, higher (49 m or 168 ft) and more commodious. In 1882 Douglass's masterpiece finished £18,745 under budget and gained him a knighthood soon afterwards. It was automated in 1982, exactly 100 years after its completion, with the addition of a helicopter landing-pad on its roof.

But the good and grateful people of Plymouth were not to forget the efforts of John Smeaton. In 1884 they had raised sufficient funds for it to be dismantled from its Eddystone pedestal and re-erected on Plymouth Hoe for all to admire. Only a stump remains at sea. So significant was Smeaton's tower that it appeared on the bronze penny coin in 1860 and – apart from a gap of 42 years – to be a constant companion alongside Britannia right up until the penny ceased to be minted in 1970. Smeaton's red and white stripe 'tree-trunk' is the

Smeaton's Eddystone lighthouse, 1759, on Plymouth Hoe

Douglass's Eddystone lighthouse, completed 1882

iconic image of a lighthouse on artwork and postcards of the British seaside to this day. His masterpiece still stands proud on the Hoe of course, extinguished now, but watching over Douglass's Eddystone still flashing on the distant horizon.

Mariners – and lighthouse keepers – obviously had a deep interest in the weather, which often determined where and when some lighthouses came to be situated in the first place. There had been storms since the beginning of time, and there will be many more. Ships and sailors had been lost since humans first pushed boats out on the water and regretfully, despite the best efforts of the rescue services and Health and Safety protocols, there will be more in the future. But, of course, few if any such tragedies had, until Defoe, been documented in quite such detail as that of November 1703.

Perhaps one of the more infamous was that of the ill-fated Spanish Armada. On 28 May 1588 some 130 ships with nearly 9,000 sailors and 20,000 soldiers set sail from Lisbon. They encountered some bad weather on the way but, as we all know, were finally confronted by Sir Francis Drake's ships off Plymouth, indeed close to Eddystone Rock. A statue to Drake stands proud beside Smeaton's tower on the Hoe. Drake engaged the ships just off the coast of Flanders. Although the English fleet outnumbered the Spanish, they were considerably outgunned. Nonetheless Drake's force prevailed. In the end only three Spanish ships were sunk and it would be subsequent events that took the toll. The surviving but battered Spanish ships – by now trapped in the southern North Sea in deteriorating weather – could do nothing but retreat by a most hazardous route – round the north of Britain. Ultimately, harassment by the English fleet, navigational deficiencies and errors (Spanish charts ended at the Moray Firth), food shortages, disease and persistent storms contributed to only 67 ships finally returning to Spain, with just over half their compliment of men. It is said that over 20,000 Spanish seamen and soldiers were lost.

El Gran Griffon – the flagship of the Armada's supply squadron – came to grief off Fair Isle in Shetland: most of the 300 on board survived by climbing the tilted mast onto the cliff! A few other Armada vessels foundered in the southern Hebrides where, for instance, the *San Juan de Silicia* sought refuge in Tobermory Bay, Mull, near where Rubha nan Gall lighthouse now stands. The local clan chief Maclean of Duart exploited the 60 crew and 270 troops as hostages in feuds against his rivals until, in highly suspicious circumstances, the

Smeaton's lighthouse on Plymouth Hoe with the current lighthouse (circled) on the distant horizon

ship was finally blown up in the harbour – and so the myth of the 'Tobermory treasure ship' was born.

The gales were relentless. In unfamiliar waters and bad weather one Spanish pilot observed how 'we sailed without knowing whither through constant fogs, storms and squalls'. Off the Irish coast between Antrim and Kerry no fewer than 24 other ships were wrecked, six in one day alone. Along one beach 1,500 drowned sailors were washed up. Back at the Spanish court, Philip II lamented how he 'had sent the Armada against men, not God's wind and waves'. Queen Elizabeth I, on the other hand, celebrated how 'England had been saved – by a Protestant wind!'

Naval tragedies tend to be better recorded and more often pass into the historical record. One of the worst took place off the Isles of Scilly only a few years after Defoe's storm. On the night of 22 October 1707 five large ships of the British fleet were lost and about 1,670 men were drowned. They were returning home from Toulon, under the command of 57-year-old Rear Admiral Sir Cloudesley Shovell in his 90-gun flagship HMS *Association*. A famous naval hero of the day, the Admiral's body came ashore on St Mary's where it was buried but later removed with all due honours to Westminster Abbey. It has been concluded that the tragedy happened because of the poor maps and navigational capability of the time, but it is strange that on such a dark, stormy night no lookout spotted the warning coal-fired beacon on the St Agnes lighthouse. Trinity House had constructed this substantial 22 m (70 ft) tower in 1680 (one of its earliest); in 1790 it was converted to oil until finally discontinued in 1911. The chauffer or cresset is probably the only coal basket to survive and can be seen in Tresco Abbey Gardens, beside a French 18-pound cannon taken as a prize by the *Association*.

It was the loss of the *Association* that led to the famous quest for a method to accurately measure longitude and ultimately to John Harrison's chronometer of 1761. The 70-gun *Eagle*, another of Shovell's fleet, sank near Bishop Rock but it would not be until 1858 that a lighthouse came to be built there – by Nicholas Douglass and his two sons, James and William. The 2,867 t four-masted *Falkland* was bound for Falmouth when on 22 June 1901 a southwesterly gale drove her at the lighthouse itself. The stays had broken and sent the vessel broadside on to the Bishop Rock. The mainyard hit the tower and the ship overturned, drowning the captain and five others.

Someone once said that 'every rock in Scilly has a shipwreck'. The islands are the first landfall at the western approaches to the narrows of the English Channel. With a full exposure to the prevailing westerlies, the open fetch stretches back across the Atlantic. So with 'many rocks terrible to behold' and a huge tidal range, it is not surprising that over 300 wrecks have been recorded around the Isles since the 17th century. More recently, the infamous 60,000 t oil tanker *Torrey Canyon* ran aground on the Seven Stones Reef – despite warnings from the manned light vessel nearby. Its 120,000 t of crude oil caused a huge environmental disaster – the worst in British waters. One hundred and ninety-three kilometres (120 miles) of the Cornish coast was contaminated (and a further 80 km or 50 miles in France), over 15,000 seabirds were killed with a further immeasurable impact on marine organisms. The only benefits to accrue subsequently were the introduction of stricter international pollution regulations.

Left: St Agnes lighthouse (now disused) in the Isles of Scilly

Inset: St Agnes coal-burning cresset and a French prize cannon from HMS Association

Waves not infrequently reach the full length of even the tallest lighthouses and, in one storm, a fishing net was found dangling from an outside aerial 70 ft (21 m) up the Bishop Rock tower. Another tempest once wrenched the fog bell from its mounts near the top. It was for this reason that, in 1882, a substantial stone apron was built around the base of Bishop Rock lighthouse, and its height raised a further 36 ft to 175 ft (53 m). Nonetheless in 1925 waves damaged the glass of the lightroom and extinguished the light. Ultimately in 1976 the construction of the helipad on top reduced the height slightly to 49 m (162 ft), but it is still the tallest lighthouse in the British Isles. (Bishop Rock is also in the *Guinness Book of Records* as the smallest island in the world with a building on it!)

St Agnes and Bishop Rock lighthouses, Isles of Scilly

Huge waves hitting Bishop Rock tower would sometimes cause so much vibration that crockery would fall off the shelves. Robert Stevenson designed the Bell Rock lighthouse (36 m or 118 ft tall and first lit in 1811) so that its masonry blocks dove-tailed both horizontally and vertically – just as with Smeaton's Eddystone. The result was so rigid that it too shuddered in big seas. On the other hand, when the second tallest lighthouse in the UK (at 48 m or 158 ft) was built in 1844 – Skerryvore, 17 km (10 miles) south of Tiree in the Inner Hebrides – Robert's son Alan calculated that the weight of the tiers built on top of each other would be sufficient to bind the whole structure just as strongly so the tower yielded slightly to the pounding waves.

Bishop Rock lighthouse, Isles of Scilly

In Britain hardly a winter passed without loss of ships, shipping and sailors. Indeed in December 1799 alone, during a tempest lasting three days, over 70 vessels foundered along the east coast of Scotland. The devastation continued into the New Year, indeed it persisted over several succeeding winters. In January 1804 HMS *York,* a 64-gun ship of the line (1,433 t) struck Bell Rock and was lost with all hands. Such tragic incidents would refuel the debate to construct a lighthouse on the Bell Rock, and Robert Stevenson himself listed 16 vessels he knew to have been wrecked there in the decade leading up to his starting the work. Its light was first exhibited on 1 February 1811 (see later) and has doubtless saved many, many lives. As have the entire suite of lighthouses that would come to illuminate our coast.

Lighthouses can never totally prevent accidents however, and are no guarantee of safety. Indeed they might even be implicated in accidents. An extraordinary event, and mercifully with no loss of life, took place in west Norway. At the entrance to Florø harbour in Sogn og Fjordane stands a tiny islet barely large enough to support its handsome wooden lighthouse. It is named Stabben because the rock rises straight up from the sea and is shaped a bit like a stump or 'stabbe'. When the station was built in 1867 all the facilities normally constructed around any other lighthouse had to be incorporated inside its cellar, with family accommodation on the floor above. (There was however just enough room for a tiny vegetable patch outside, which had to be removed every autumn to prevent the soil being washed away.) In 1899 a ship hit the building and its bow pierced through the wall of the accommodation where the keeper and his family were eating a meal! Despite this, it was not until 1975 that this lighthouse was finally automated and is now rented out as holiday accommodation.

Great Britain lies at the northwestern extremity of Europe, exposed to the full fetch of the Atlantic, with the North Sea as its eastern edge. Waves and currents are funnelled through the English Channel. Tides range from a metre or two to 13 m (42.7 ft) in the Severn Estuary, second only to the 15 m (49.2 ft) in the Bay of Fundy on the other side of the Atlantic.

Lighthouse on Inisheer, Aran Islands, Eire

Stabbe lighthouse, Florø, Norway

The ultimate factor determining the perilous nature of our coastal seas is weather. The climate of the British Isles is highly variable and all four seasons can be experienced in a single day. So it is not surprising that weather is a national obsession. People living on the coast are preoccupied by it in ways that landlubbers are not. So too are mariners of course, with their particular 'weather eye': it is not so much rain or sun that concern them but wind. And, from the brief discussion above, the impact of strong winds – all too often with dire consequences – must already be apparent.

The climate of the British Isles, in particular, is a consequence of their location. The country may be small but it extends south to north from 49 to 60 degrees north offering some gradient of temperature (and day length). Depending upon the course of the jet stream in the upper atmosphere, our prevailing winds are southwesterly but cold air may sometimes blow down from the north or, at other times, warmer air from the south. Our islands lie on the northwestern edge of the Eurasian continent – the world's largest land mass – so in addition, receive dry continental air masses from the east, usually warm in the summer, often frigid in winter. It is constant conflict between all these air masses that give us such variable, unsettled and often violent weather.

And even within our small country there are significant differences. The east coast is cooler and drier while moist, mild air rolling in from the Atlantic diminish temperature extremes in the west to produce cool summers and mild winters. But the west coast is famously wet and windy. To the west our prevailing winds can blow 3,000 miles (4,800 km) over the North Atlantic Ocean while in contrast there is at most 350 miles (563 km) between UK and Scandinavia. True, the North Sea can be whipped into a fury but it is generally accepted that eastern shores of Britain, with much less of a fetch, are less likely – most of the time – to generate mighty seas. With his professional interest in lighthouses, and more than a passing interest in meteorology, Thomas Stevenson – Robert's youngest son and Robert Louis the writer's father was of the opinion that storms are usually less severe in the North Sea than in the Atlantic. With the ever-enquiring mind so typical of all his family, Thomas recorded, for instance, a force of nearly three tons per square foot at Skerryvore during a heavy westerly gale on 29 March 1845. (Amongst a host of innovations in engineering and optics, he was also responsible for the invention of the white louvered 'Stevenson screen' so long employed in meteorological recording).

Autumn is particularly prone to unsettled weather. Cool polar air moves southwards often to meet warmer air from the tropics directly over the British Isles, producing an area of great turbulence. Combined with an ocean warmed over the spring and summer the unsettled weather of autumn is produced. Cool air over the land reacting with a warmer ocean also brings rain. Atlantic depressions during this time can become intense and winds of hurricane force (greater than 119 km/h or 74 mph) can be recorded. Western areas, adjacent to the Atlantic, experience these severe conditions to a significantly greater extent than eastern areas, especially so in the autumn months and persisting into early winter, often the wettest and windiest time of the year.

From 1868 Scottish lightkeepers were charged with taking meteorological records twice a day. They also recorded wind, thunder, even auroras 'with an accuracy and fullness

Stormy seas

attempted nowhere else': the following year they added fog to their observations. On Bell Rock John Maclean Campbell noted: 'In snowstorms such as we have had of late, our lantern soon becomes plastered up with snow on the weather side, necessitating constant removal to prevent it from completely blinding our light in that direction. This is an operation often accomplished with difficulty, especially when carried out in the teeth of a gale.'

The north and west of Scotland is much windier than the south of England. Annual average wind speeds can be up to three times higher in Shetland than the Isle of Wight. Gales are defined as winds of 51–101 km/h (32–63 mph) and are strongly associated with the passage of deep depressions across the country. The Hebrides experience on average 35 days of gale a year while inland areas in England and Wales receive fewer than five. During the night of 12 January 2005, my home in Uist in the Outer Hebrides was battered by one of the worst hurricanes in living memory. For six hours we experienced constant winds (not gusts) over Beaufort Force 12: Benbecula Airport recorded 140 mph (225 kph) winds from the southwest. Extreme high spring tides combined with a considerable storm surge from the open Atlantic caused massive erosion and flooding while, most tragic of all, five members of the same family – three generations – were swept to their death in two

Old print of Eddystone light-room after a snowstorm.
Strand Magazine, *1892*

cars. At least two houses were blown down, most others suffered some degree of damage, and in places fences, draped in seaweed blown off the shore, were ripped up and draped across the top of telegraph poles; but all this faded into insignificance against the appalling human tragedy.

Mountain-tops, as well as exposed northern coasts, are the windiest situations in the country. I myself was on the slopes of Cairngorm Mountain on 20 March 1986 when the summit recorded a gust of 173 mph (278 kph). However, I recall, the anemometer was read manually, with no recording or printout to back up the observation so, at the time, the Meteorological Office was slow to accept it. Shetland held the British wind speed record of 177 mph (285 kph) near Muckle Flugga lighthouse in 1962. Then again, on 19 December 2014, the anemometer stationed close to the summit of Cairngorm Mountain registered a wind speed of 194mph (312 kph) (only 37mph short of the world's most powerful gust on Mount Washington in New Hampshire in 1934). But once again there was no proof of the observation so the previous record of 177 mph still stands.

three

The Dark Coast

Nowhere does the sea hold wider sway. It carries to and fro a mass of currents, and in its ebb and flow is not restricted to the coast, but passes deep inland and winds about, pushing in even among highlands and mountains as if in its own domain ... The climate is foul with frequent rains and mists, but there is no extreme cold. Their day is longer than in our part of the world. The night is bright and, in the furthest part of Britannia, so short that you can barely distinguish the evening from the morning twilight.

Tacitus, circa AD 98

The Romans were responsible for Britain's first lighthouses, only one of which survives. It stands, reasonably well preserved, in the grounds the 12th century Norman castle of Dover, the scant remains of another lie on the opposing Western Heights across the town. They date from the reign of Emperor Claudius around AD 46–50, some three years after the invasion of Britain by Julius Caesar and three decades before Agricola, the great Roman general and governor of Britain. The one in Dover Castle is now only a four-storey building 19 m (62 ft) tall, the top floor being a medieval restoration when it was used as a bell tower. Originally it may have been six or eight storeys. A beacon of fire would have burned on top every night marking the harbour entrance for Roman ships crossing from Gaul. It is apparently modelled on Emperor Caligula's earlier lighthouse on the opposite side of the English Channel near Calais, built in AD 40. In classical times, at least a couple of dozen lighthouses may have lit the coasts of the Mediterranean, especially of Greece and Italy, so the concept of illuminated navigational aids was already well established by the time the Channel towers were erected. A detailed account of early lighthouses can be found in D Alan Stevenson's 1959 book *The World's Lighthouses from Ancient Times to 1820*.

Around 800 BC – nearly a millennium before the Roman towers – Homer wrote of fires being lit to guide seafarers at night. Another Greek writer, Leches, writing about 660 BC, tells of a fire tower burning at the entrance to the Dardanelles – presumably built by the Phoenicians who regularly traded throughout the eastern Mediterranean. One of the

Seven Wonders of the Ancient World – the Colossus of Rhodes – may have been built as a lighthouse as long ago as 3000 BC. Another, the 465 ft (142 m) tall Pharos, was constructed as a lighthouse between 283 and 247 BC to warn sailors of the treacherous sandbars off Alexandria, one of the busiest ports of the ancient world. It consisted of a three-stage tower of white marble, decorated with sculptures of Greek deities and mythical creatures. On top blazed a beacon of fire, perhaps focused by mirrors of polished bronze, and was said to have been visible for 30 miles (48 km) or more. 'Soaring in height, a work of unimaginable construction' according to Caesar, Pharos remained among the tallest of man-made structures until the Eiffel Tower was built in 1889. During the complex love-plot between Caesar, Mark Anthony, Cleopatra and her family, Caesar stormed the strategic Pharos, an image of which ended up on the tomb of Arsinoe, Cleopatra's murdered sister, at the Temple of Artemis in Ephesus, Turkey – another of the seven wonders of the ancient world.

The Pharos lighthouse was still functioning when the Arabs conquered Alexandria in AD 642, but an earthquake damaged the lantern about 50 years later; and again in 1303. By 1349 it was in ruins, but divers have recently located what seem to be its remains on the seabed near the harbour. The name – which actually came from the island upon which it stood – lives on; 'pharology' is the study of lighthouses while, in 1807, one of the NLB's first ships was named *Pharos*. The tenth and current vessel of the same name came into service exactly 200 years later.

Northern Lighthouse Board's Pharos, *the tenth of that name*

The Dark Coast, Isle of Mull

As we now know, Agricola's Roman fleet was not the first to circumnavigate the British Isles. Pytheas of Massalia, a geographer and explorer from the Greek colony at Marseilles, undertook a remarkable voyage in northwest Europe sometime around 325 BC. He 'travelled over the whole of Britain that was accessible'. Pytheas probably utilised local knowledge and whatever vessels and crew he could charter, but he was boldly sailing dangerous waters, hitherto unknown to the 'civilised' world, so knew not what to expect.

Unfortunately no account of this incredible venture survives, only fragments mentioned – somewhat dismissively – in a 17 volume *Geography* written around AD 20 or so by a later Greek traveller called Strabo. Others, equally unconvinced of Pytheas's veracity, nonetheless freely pillaged his writings. They include the second century BC Greek astronomer Hipparchus, Pliny the Elder in his *Natural History* (first century AD) and Ptolemy in his *Geographia* (second century AD). For six days, we learn from these accounts, Pytheas even ventured northwest of Britain into the Arctic until 'an ocean of slush ice and fog that one could not sail through it' stopped him. In addition Pytheas is credited with the proposal that the moon influenced tides.

In AD 150 Ptolemy produced the first map of Britain – although he depicted Scotland with a 90° error in orientation (which in fact would be perpetuated in maps and charts for many centuries to come). Ptolemy did locate quite a few physical features. The exact identity of some of them are still open to debate but, in the Hebrides, 'Scitis' for instance is obviously the Isle of Skye, 'Malaius' Mull and 'Botis' Bute. 'Dumna' was once thought to refer to the Isle of Rum but is in all likelihood the Isle of Lewis.

It is indeed remarkable to think on these classical ships sailing cautiously round the complex and convoluted coast of Britain. They had no idea what to expect nor, of course, did they have any maps or charts. Their voyages are reminiscent of Captain Cook and other early navigators tentatively picking their way through uncharted waters beyond the furthest limits of the known world or even – in modern times – of the Apollo astronauts boldly going to the moon where no man had gone before.

The Icelandic sagas tell how in AD 869 a Viking, Floke Vilgerdsson, built a fire cairn near his home port to guide himself back to Norway; he also took three ravens to use, much as Noah did, to find land. Apparently this stone beacon in Norway was maintained for 1,000 years until local herring fishermen demanded a proper lighthouse be built instead. Around 1431 an Italian seafarer Querini journeyed along the Norwegian coast using 'beacons on the tops of the islands which thereby indicated the best and deepest passage'; these did not however prevent him from running aground in Lofoten. The first proper lighthouse in Norway seems to have been established at Lindesnes, at Norway's southernmost point, in 1656. Ultimately about 150 manned lights and many more unmanned beacons came to be established along the Norwegian coast and all had been automated by December 2006.

France also has a long coastline with a proud lighthouse tradition and no account of lighthouses would be complete without mention of Cordouan. The Bay of Biscay, with deep troughs in its seabed, is notorious for turbulence and the Gulf of Gascony can be particularly inhospitable in a northwesterly gale. The wide mouth of the Gironde is fully exposed but is the point of access to the port of Bordeaux, France's main outlet for wine. Dangerous rocks were marked since AD 880 when King Louis the Pious (814–840), the son of Charlemagne, permitted the citizens of Bordeaux to build a small beacon on one particular islet.

The first proper lighthouse at Cordouan was built for Edward, the Black Prince (1330–76) when, in the 14th century, Gascony was under the English throne. Although nothing remains of it now, it seems to have been a tower 15 m (50 ft) high topped by a chauffeur or brazier containing a fire, which was tended by a hermit monk. There was a small chapel alongside and a small community of fishermen and pilots grew up around it. Although now 7 km (4.5 miles) offshore, it seems back then that the islet may well have been attached to the coast. After the monk died however, the beacon fell into disrepair.

In 1584 an eminent Paris architect Louis de Foix was charged by King Henry (1574–89) to build a new lighthouse but before it was completed in 1611 both men were dead and King Louis XIII had ascended the throne of all France. Recent commentators have variously judged that the structure was self-indulgent and far too grand for a lighthouse, also that it possessed 'exuberant external decoration'. (Both descriptions could equally apply to Winstanley's later Eddystone efforts.) Cordouan was a Renaissance amalgam of royal palace, cathedral and fort. The base was a circular stone wall 8 ft (2.4 m) high and 135 ft (41 m) in diameter to absorb the force of the waves. The tower had four storeys, each a profusion of gilt, carvings, statues and elegantly arched doorways. The ground floor held accommodation for four lightkeepers, above was a sumptuous apartment for the King, with a chapel on the third floor, a secondary lantern on the fourth floor and the blazing beacon itself on top, 162 ft (48 m) above the sea.

In 1782 the first oil lamp with parabolic reflectors replaced the fire brazier, but by then Cordouan was in poor repair and deemed not high enough. Eight years later everything above the chapel was removed so that the lighthouse could be extended to an unprecedented height of 197 ft (60 m) to incorporate an innovative revolving Argand reflecting lamp, fuelled by whale, olive and rapeseed oils. By this time the forces of erosion had turned the building into a rock station in the true sense, effectively protected by its seawall. Cordouan was vandalised shortly afterwards during the French Revolution but it was soon repaired, and Augustin Fresnel frequently used it for experiments in optics. In 1822 he installed the first permanent Fresnel rotating lens system.

On a visit to France in 1834 Alan Stevenson – who was greatly influenced by his friend Fresnel's lenses – found Cordouan decaying, and its stairway running with water because walls and windows leaked. The building became a historic monument in 1852 – the same year as Notre Dame in Paris – and has been restored and modernised several times since. It remains one of the oldest lighthouses still in operation but now only the tenth tallest in the world. Today it is considered the doyen of them all and the patriarchal Pharos of Europe.

In French the word *phare* is reserved for larger, coastal lighthouses, smaller and harbour lights being called a *feu* – literally fire. For many years all navigational aids came under the regulation of the Bureau des Phares et Balises but now this venerable body, with some 120 lighthouses, has been absorbed into Signalisation Maritime, within the Ministère de l'Écologie, du Développement Durable, et de l'Energie. Dues are no longer collected in the French system, with many lights actually operated by transport ministries or port authorities.

French ports face the Atlantic or the English Channel so the country has long played a significant role in lighthouse development. But it has taken many centuries for lighthouses, modern charts, compasses, buoys and GPS to come into widespread use, so we must marvel at early mariners, picking their way cautiously between rocks, sandbanks, mist, tides and currents, with nothing but sails and – in smaller vessels – oars, to keep them out of trouble. Further out to sea was safer but approaching or leaving harbour demanded particular skill and courage.

Local knowledge was and is always a vital ingredient, hence the continuing demand for pilots. In 1679, after 16 hours at sea and with a storm brewing, a Skye gentleman Martin Martin only reached St Kilda – 40 miles (65 km) from land (the Monach Isles) – because his crew knew to follow the seabirds flying towards the archipelago's magnificent cliff colonies. Nearly two centuries later it was still said how the sound of seabirds reflecting off the cliffs at South Stack near Holyhead helped orientate mariners in foggy weather, not just back in the silent days of sail but also during the last war to the crews of the pilot boats.

For a long time mariners depended upon experience, or memorised sailing instructions until, around 1540, a pilot Alexander Lindsay wrote them down for James V of Scotland. The King wished to sail his fleet round the north of his realm to subdue the rebellious Hebridean lords. Lindsay included in his 'rutter' (from the French *routier*) the course of tides, the time of flood and ebb, the appearance of the coastline with conspicuous landmarks, the headings

Tall ship off the coast of Mull

and distance from one headland to another, havens, roads, sounds and other dangers. Essential equipment included a magnetic compass, a sandglass for timings and a lead line for soundings. Although none of the surviving copies of Lindsay's rutters now include a map, one published soon after and credited to him was 'a much better outline of Scotland than any previous map … in fact more accurate than any other later maps of the seventeenth century'.

In England, maps and charts – of varying quality and usefulness – began to be produced in the 16th century, and many of the likely cartographers were Scotsmen in Henry VIII's court, and who probably had some connection with Lindsay. One of them was John Elder from Caithness, working from 1533–65. Then in 1686 the Scottish Parliament commissioned John Adair (1660–1789) to prepare charts, which were published in 1703. At the same time, south of the border, the Admiralty gave funds to Trinity House for Captain Greenvile Collins to chart the British coast from 1681–8, often incorporating existing charts and seamen's sketches and notes, as well as his own observations – 'a most exact and useful undertaking'.

The Scottish and English parliaments united in 1707. Soon afterwards a naval commander from Orkney, Murdoch Mackenzie, began surveying the north of Scotland and in

*South Stack lighthouse,
Holyhead, North Wales*

1750 published eight charts of Orkney and Lewis, leading to his being commissioned by the Admiralty to survey the whole west coast of Britain and Ireland. His efforts were a great improvement on existing charts. But not until 1795 with the founding of the Admiralty Hydrographic Office – under another Scotsman Alexander Dalrymple – did charts of the British coast really become relied upon for safe passage, at least by daylight. England had already established some daymarks along the coast as aids to navigation, but, until lighthouses began to be constructed, night voyages close inshore were still highly dangerous.

Maps and charts are now commonplace and illustrate just how – for all their limited extent – the British Isles and their coasts are remarkable for variety and richness of landscapes. Much has now been modified by humans of course – 'painted by man on a landscape provided by geology' as someone once put it. Some parts of the coast are impacted less by man, particularly in the north and west. Nonetheless the location of lighthouses around Britain reflects this diversity of topography – atop dramatic cliffs on the oldest of our rocks, below white cliffs of some of our youngest, on isolated wave-washed skerries off rocky shore, or along sandy beaches and estuaries.

Butt of Lewis lighthouse, on Lewisian Gneiss, Outer Hebrides

The Needles lighthouse on chalk, Isle of Wight

Geologically we can subdivide Great Britain into a Highland or Atlantic Zone (to the north and west and including the southwest of England) and a Lowland or North Sea/Channel Zone mainly to the south and east. There is a vaguely similar trend within Scotland itself. Low, sandy shores and wide firths stretch along its own east coast while the northwest, where Cape Wrath and the Butt of Lewis lighthouses came to be built for instance, has the oldest rocks in Britain: this Lewisian gneiss is up to three billion years old. Slightly younger cliffs of Torridonian sandstone, upon which Stoer Head light now stands, are also of the same Precambrian age. At that time Scotland lay off Antarctica, and life consisted of little else but single-celled bacteria. The gneiss is harder and more resistant to erosion, reflected in the spectacular cliffs of Cape Wrath and the Flannans, for example. Devon, Cornwall and the Isles of Scilly are less ancient – granites, grits and sandstones – but lower lying.

Cape Wrath lighthouse, northwest Sutherland in a storm

Stoer Head lighthouse, Sutherland, with Canisp and Suilven behind

The south of Britain was not affected by glaciation to the same extent as the north, even at the glacial maximum 450,000 years ago. Thus, in complete contrast, the lowlands south of a line between the Bristol Channel and the Wash are softer, younger clays, sand, sandstones, limestones and chalk often of Jurassic Age, only 200–142 million years old. The Old Red Sandstone cliffs of Orkney and Caithness date from 350 million years ago, at a time when Scotland, England and Wales were united, crossing the Equator during their great tectonic journey across the globe.

Durdle Door, Dorset's Jurassic Coast

The cliffs of Hoy, Orkney with the Old Man to the right

Nowadays, the British Isles comprise – by some assessments – some 6,000 islands around a coastline stretching for 800 miles (1290 km) from north to south: and a maximum of 500 miles (800 km) east to west. Just as it is difficult to judge what defines an 'island' however, it is hard to find consistency for the documented length of Britain's coastline. It all depends upon the resolution one chooses. If one uses a metre-stick to mark out the circumference of a large circle on the ground for instance, one will come up with a lesser figure than if one had used a 30 cm ruler. Using 1:10,000 maps at mean high water mark, the Ordnance Survey quotes the mainland of Britain together with its larger islands at 19,491 miles (31,368 km). Of this, England (including Lundy, the Isles of Scilly and the Isle of Wight) contributes 6,261 miles (10,077 km), Wales (including Holy Island and Anglesey) 1,680 miles (2,740 km) while Scotland (including Arran, Islay, Jura, Skye, the Outer Hebrides, Orkney and Shetland) extends no less than 11,550 miles (18,588 km). Scottish Natural Heritage (in *Scotland's Living Coastline*, 1999) uses slightly different figures but offers some helpful comparisons. The UK is 2.4% of Europe's land surface yet holds 13% of its coastline (England 3.8%, Wales 1.1% and Scotland 8.2%).

As recently as 20,000 years ago an ice sheet a mile thick covered Britain, right down to a line between the Severn and Humber estuaries. As the climate began to ameliorate and the ice slowly melted, Britain emerged from an Arctic environment to become more amenable to life. It is not that long ago that Britain was still part of the whole European land mass. Prior to 6000 BC – Mesolithic times – the English Channel did not exist: even much of the southern North Sea was dry land. Plants and animals could easily encroach overland until the seas rose, swelled with meltwater, to flood the English Channel and Dogger Land. Relieved of its heavy burden of ice, the land also rose, but at a slower pace, and indeed in places it is still rising. However, with the prodigious ice melt, the rising sea was to outpace the land, so inevitably the Channel was created and Britain became an island. But it would still

be several thousand years more before Agricola's fleet of ships revealed this to geographers for the first time!

To reach the island of Britain, plants and animals had to overcome a sea barrier: by flying (birds and bats, many insects and other invertebrates), by swimming or drifting (for animals perhaps on rafts of vegetation), by stowing away on boats or even by being introduced deliberately by people. The very first humans entering Britain probably arrived overland, before the sea cut it off. Yet more arrived later by hugging the Atlantic coast in dugout canoes (a remarkable example is to be found in Dover Museum) or flimsy skin- or bark-covered boats. They probably travelled only by day finding overnight shelter in numerous inlets, coves and estuaries, settling in for the winter before some opted to resume their voyage as soon as the weather improved again. They may have had an ability to navigate by the stars but more than likely they were content to fix on cues visible on a distant horizon, before striking out in the hope of reaching landfall before nightfall. In such frail craft, they not only colonised Britain but also reached the numerous islands of its west and north. A few brave souls even reached St Kilda over 40 miles (65 km) out into the Atlantic.

Three million of the 64 million people in Britain nowadays still live on our coast, its hinterland or up river valleys, and for many boat travel is still important. Indeed, nowhere in Britain is more than 70 miles (113 km) from the sea. Only 290 of Britain's islands are permanently inhabited. Scotland alone has some 800 islands, 99 of which are populated by up a handful up to 100,000 folk: only 14 have over 1,000 people and 40 have fewer than 100. The British Isles are very much a nation of 'islanders' and, since Tudor times, they were to become a major maritime force across the globe.

From its coast Britain could reach out to the world but so, too, could the world reach in. Indeed, along with France, our coastline commands the western approaches to all the major ports of Northern Europe. Easy access for trade also means it is vulnerable to attack from the sea. In commissioning elaborate coastal defences, Henry VIII wished to redefine

England, no longer as part of Europe but as separate from it – a nation apart. However, the needless loss of his now famous flagship, the *Mary Rose*, demonstrated how England was not yet a naval superpower. Henry's daughter Elizabeth I came to be a significant monarch in this respect, and presided over a famous naval victory against the Armada. Yet there was still a profit motive behind her ships: the most famous of her seamen Sir Francis Drake was, at the end of the day, still a privateer – ostensibly a pirate, like many others of his time, sailing under the favour of the English flag. By the time Charles I took the throne of a now united Britain, its maritime power came to be reinforced. 'He called his great warship *Sovereign of the Seas* – a statement of intent,' as one historian put it.

Since then our history positively bristles with naval heroes and sea battles. Certainly, it was important to defend the coastline, but it was now obvious to exploit the good harbours to promote overseas trade and further develop the economy. From the 15th century, trade with Europe and beyond through English ports was buoyant. The stage was set for the development and formalisation of yet more lighthouses.

Hitherto the planting of lighthouses tended to be somewhat haphazard depending on an element of strategy, yes, to mark the most important hazards on particularly busy sea routes or the approaches to important harbours, but also to some extent on an economic whim of landowners sensing a fast buck. There would be no particular plan or programme behind the addition of new lights until our various lighthouse authorities came into being – Trinity House for England and Wales, the Commissioners of Northern Lighthouses (or the Northern Lighthouse Board, NLB) for Scotland and the Isle of Man, and the Commissioners of Irish Lights (CIL) for Ireland. By the year 1800, with only 135 lighthouses in the whole world, the English coast had become the best lit in Europe (with France and its 20 lighthouses not far behind). On the other hand in 1800, not only was the perimeter of Scotland poorly lit but also parts were still rather poorly surveyed.

In rectifying this regrettable situation the name Stevenson looms large in 19th century Scotland. English and French lighthouse engineers certainly knew their business and all seem to have shared information and ideas freely across borders. But several generations of Stevensons proved particularly prolific and contributed detailed analyses of lighthouse construction to many venerable institutions, encyclopædias, periodicals and journals. It is not appropriate here to detail technicalities, only the natural factors to be considered in both the location of a lighthouse and its ultimate design. Alan Stevenson, in particular, drew together and published a useful set of 'guidelines, which were developed further by his brothers and nephews. The numbering adopted below is his. Some are fairly obvious and others are very technical, but it is worth mentioning a few that are particularly relevant to natural features or phenomena, our theme in this book.

1. The most prominent points along a line of coast, or those first encountered on over-sea voyages, should first be lighted; and the most powerful lights should be adapted to them, so that they may be discovered by the mariner as early as possible before his reaching land.

It is interesting to note that of the 90 or so major light station names around Scotland, nearly one half (40) include the words 'head', 'point', 'ness', 'mull', 'butt', 'cape' or 'rubha' (in any of its Gaelic spellings), revealing their situation at prominent points along the coast. Furthermore two thirds of the 90 are on islands, the rest along the Scottish mainland.

2. So far as is consistent with a due attention to distinction, revolving lights of some description, which are necessarily more powerful than fixed lights, should be employed at the outposts on a line of coast.

7. Lights intended to guard vessels from reefs, shoals or other dangers, should in every case where it is practicable be placed seaward of the danger itself, as it is desirable that seamen be enabled to make the lights with confidence.

8. Views of economy in the first cost of a lighthouse should never be permitted to interfere with placing it in the best possible position; and when funds are deficient, it will generally be found that the wise course is to delay the work until a sum shall have been obtained sufficient for the erection of the lighthouse on the best site.

9. The elevation of the lantern above the sea should not, if possible, for sea-lights, exceed 200 ft; and about 150 ft is sufficient, under almost any circumstances, to give the range which is required. Lights placed on high headlands are subject to be frequently wrapped in fog, and are often thereby rendered useless, at times when lights on a level might be perfectly efficient. But this rule must not, and indeed cannot be strictly followed, especially on the British coast where many projecting cliffs would obscure lights placed on adjacent lower land.

11. It may be held as a general maxim that the fewer lights that can be employed in the illumination of a coast the better, not only on the score of economy but also of real efficiency. Too many will confuse the mariner.

Alan's younger brother, Thomas, went on to consider more technical recommendations for rock lighthouses, based on his particular studies of wave action and impacts. Without doubt the Stevensons, alongside the Douglasses and lesser dynasties or individuals of marine engineers – all standing on the shoulders of John Smeaton – amassed a phenomenal amount of knowledge and experience, which they were only too willing to share with the world – *in salutem omnium*, for the safety of all, as the motto of both the NLB and CIL emphasises.

As we shall see in the following chapters, Alan's point nine was probably heavily influenced by three particular light stations planted by his father Robert – Cape Wrath, the Calf of Man and Barra Head, all too often obscured by fog. (At 208 m or 692 feet above the sea Barra Head is the highest of any British lighthouse.) The same problem occurred with the first towers at The Needles (built 1785) and Beachy Head (1828), necessitating both to be replaced (in 1859 and 1900–2 respectively). It was exactly this problem that also faced the builders, during the First World War, of the unconventional West lighthouse on Rathlin

Calf of Man lighthouses in fog

Island, 5 miles (8 km) off the Antrim coast. A red light was chosen since it is known better to penetrate foggy conditions. Due to its elevated location and the need to avoid the light being obscured by fog, the tower had to be built almost a third of the way down the cliff face. Furthermore the lantern is located at the *base* of the tower, with the keepers' quarters above. Only the Irish would have the ingenuity, courage and wit, to build a lighthouse upside down!

Above and opposite: Rathlin Island's west lighthouse, County Antrim, Northern Ireland

four

Planting the Tower

… in determining the best position for a lighthouse the engineer must generally adopt the most exposed and inhospitable bit of land he can find, and there he plants his tower, defying the elements, and despising the shelter which all other mortals seek in fixing their abode.

David Stevenson, 1864

Up until the medieval times, 30 to 40 lighthouses are known to have existed in the British Isles, most of them on the south coast of England and most maintained by monks and monasteries. Monks on the Monach Isles for instance, 6 km (4 miles) west of North Uist, are said to have maintained a beacon in the fifth century, perhaps the oldest known in Scotland. The current lighthouse stands on the westernmost and smallest island of Shillay. There is also a tradition that St Gerardine displayed a lantern on the Moray coast near the present Covesea Skerries lighthouse, and he is portrayed on the coat of arms of the nearby village of Lossiemouth.

But the oldest lighthouse still operational in Great Britain and Ireland is reckoned to be Hook Head in County Wexford, perhaps even one of the four oldest in the world. A Welsh monk called Dubhán established it in the fifth century as a cell and oratory. He chose the place for its remoteness: Rinn Dubhain translates from the Irish for 'fish hook' hence the name Hook Head. Soon the monk began to find distressed mariners and wreckage on the beaches, saving as many lives as he could and burying the dead. So he constructed a chauffer or cresset, a huge iron basket to hang over the cliff. At night he kept a fire burning within it, using coal, wood, charcoal and of course driftwood from the wrecks. His fellow monks continued the tradition well after the Normans built a fortified watchtower, which is still in good condition 800 years later.

The monks departed in the 16th century, just prior to the dissolution of the monasteries in Ireland, but the light was maintained. The assistant keeper lived on the second floor, the principal lightkeeper on the third. The area became a focus for nefarious, illegal activities until the lighthouse was abandoned towards the end of the 17th century. By petition it was

Aerial view of the Monach Isles with the lighthouse on Shillay (Tony Mainwood)

refurbished, heightened and relit around 1677. The present whitewashed tower, with two black bands, is 35 m (115 ft) tall and was automated in 1977, and now includes a heritage centre. Interestingly, in 1170 an invading force under the Earl of Pembroke (known as Strongbow) was instructed to take Waterford 'by Hook or by Crook' – the latter another suitable landing place on the other side of the river.

To the west of Hook Head, in neighbouring County Cork, a lighthouse was built in 1202 on the cliff just to the west of Youghal harbour (and now virtually on the main street). The original tower was 7.5 m (24 ft) tall and 3 m (10 ft) in diameter. It became a condition that the local nuns should maintain the light, which they did until the reformation in the 1530s, when the lighthouse and the convent were confiscated. The beacon was discontinued *c.*1542. George Halpin designed the current lighthouse on the same spot, made from granite imported from Scotland; work was completed in February 1852 and it has a lantern 24 m (79 ft) above sea level. Interestingly it is said that in 1588, when he was Lord Mayor of the town, Sir Walter Raleigh planted the first potatoes in Europe in a field near the lighthouse.

A beacon was exhibited from the top of St Michael's Mount in Cornwall in the middle of the 14th century. St Catherine's Oratory on the Isle of Wight came to be built by a rich

Clockwise from top left:
Hook Head lighthouse, County Wexford, Eire
St Michael's Mount, Cornwall, looking back to Marazion on the mainland
Coquet Island, Northumberland (Ian Baker)
Eider duck incubating

merchant desperate to avoid excommunication for buying wine from wreckers; it shone from 1314 to the Dissolution of the Monasteries in 1534.

Another of many other such 'ecclesiastical lights' was on Coquet, a small island about a mile offshore, near Amble on the Northumberland coast. There is no actual record of monks maintaining a beacon here, but when a lighthouse came to be established in 1841, it was built upon the ruins of an old Benedictine Abbey founded as long ago as 684. Its very first keeper was William Darling, elder brother of the fabled heroine Grace. It was while visiting William on Coquet in the summer of 1842 that Grace contracted a chill, which developed into tuberculosis and brought about her death a few months later.

Coquet Island comprises around 6 ha of flat grassy, maritime sward surrounded by rocks. As an RSPB Reserve it is a European Special Protection Area for one of our rarest seabirds, the roseate tern. Some 90% of the UK population nests among thousands of common, Arctic and Sandwich terns. On this tiny island there are also several thousand puffins, with a few hundred breeding eider ducks. The RSPB wardens live in the former lighthouse accommodation.

To the north of Coquet Island, the Farnes lie 2.5–7.5 km (2–5 miles) offshore, more than a dozen low islands with basaltic columns of igneous dolerite forming cliffs *c.*20 m high, and owned by the National Trust. Like Coquet, the Farnes hold over 35,000 breeding pairs of puffins, nearly 2,000 pairs of terns (few if any roseates), nearly 1,000 pairs of shags, but also some 50,000 individual guillemots, maybe 4,000 kittiwake pairs with lesser numbers of razorbills, fulmars, cormorants and other seabirds. The Farnes are justifiably famous for up to 6,000 grey seals that assemble each autumn to give birth to several hundred silky white pups. But the most celebrated creature, forever associated with the 7th century hermit St Cuthbert, are the eiders – known locally as Cuthbert's or Cuddy's ducks, over 400 pairs of them. He protected the ducks, along with the other seabirds of the islands, probably the first bird protection measures ever.

In maritime history, the Farnes are renowned for their lighthouses, and for one in particular – Longstone. Monks may well have maintained beacons on the Farnes from as early as the 9th century but in 1778 two lighthouses were established, one on Inner Farne near the site of St Cuthbert's hermitage. The other on Staple Island soon had to be relocated – in 1791 – to Brownsman Island further out. The keeper Robert Darling was also relocated. His eldest son William served as boatman to the Farnes' lighthouses until, shortly after

Inner Farne, Northumberland, lighthouse and chapel

his marriage in 1805, he was appointed assistant keeper to his father on Brownsman. He kept meticulous records in his journal of birds he had seen or shot on the islands, while also corresponding with many eminent local naturalists. On Robert's death in 1815 William became principal lightkeeper; his eldest son – also William – was born in 1806, and would later become a keeper on Coquet Island. William's seventh child and fourth daughter, Grace, was born later that same year: twins George and William Brooks were born in 1819. This station was again moved in 1826 – together with the Darlings – to a new light on Longstone, incorporating the first revolving, flashing optic in the world. William's salary from Trinity House was £70 per annum plus a bonus of £10 for satisfactory service. The family served as unpaid assistants in return for their board and lodgings. At first they had also been able to supplement their income with paying guests, for the Farnes were already an attraction to naturalists. Indeed, with his interest in wildlife, William would laboriously move large quantities of sand to create new bird nesting habitat. Longstone however was bleak, offering little chance of keeping a garden or any animals. In fact he or his sons would row daily back to Brownsman to keep their garden and livestock there. But, of course, the surrounding waters were exposed to the full fury of the North Sea.

Farne Islands, old lighthouse on the Brownsman (Ian Baker)

At 4 am on 7 September 1838 the steamship *Forfarshire*, on passage from Hull to Dundee, struck rocks off the Longstone reef. At 4.45 am Grace looked out her bedroom window to see the wreck. It was not until 7 am that the family saw any sign of movement from survivors. Her father deemed that it was far too rough for lifeboats to come from the mainland and, with his 18-year-old younger son William Brooks off the island, William senior immediately set off – with Grace – in their 6 m (20 ft) rowing boat. They doubled the actual distance they had to row by seeking lee round the island. Grace held the boat steady while her father leapt ashore to help the survivors. They could only take five – a woman, an injured man and three others. Two of them, both crewmen, were required to assist rowing back to safety against the tide; and then to return a second time with William to lift off a final total of nine altogether – four passengers and five crew – from a complement of 62. A further nine survivors were later picked up far to the south in their drifting lifeboat. In the meantime William Brooks, Grace's brother, and six lifeboat men from Sunderland had arrived at the lighthouse, where they were then marooned by the weather for two days. Within weeks of the rescue Grace had become a national heroine and a national subscription raised a substantial reward for her, with lesser sums for her father and the Sunderland lifeboat. She spent a further three years with her parents on Longstone, where she became somewhat of a tourist attraction, before her health began to deteriorate and in 1842 she died, aged only 26.

Longstone lighthouse, Farne Islands, Northumberland (Ian Baker)

In 1826 Trinity House enlarged Longstone lighthouse to the present red and white circular tower, with much improved optics. It was automated in 1990 eventually to be monitored and controlled from Trinity House's principal depot in Harwich, Essex. In June 2005 the National Trust purchased the lighthouse cottages for their wardens and the property is a highly popular tourist attraction.

Apart from various lighthouses maintained by the church, others came to be constructed in forts and at harbour entrances. Trade with Europe and beyond was buoyant in the English ports from the 15th century. Furthermore ships crossing the Atlantic and approaching landfall needed to orientate themselves to ports in the southwest of Britain and the English Channel, so appropriate beacons were proving a useful aid to navigation. Private individuals took an interest in constructing lighthouses at strategic points along the coast, since funds could be derived from tolls levied from passing ships when they entered harbour. In some cases this proved a remarkably lucrative business.

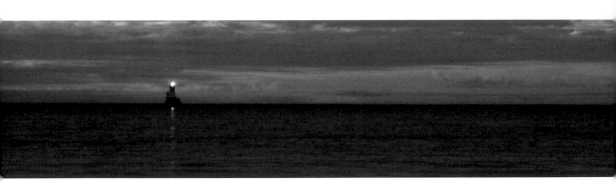

Longstone lighthouse at sunset

Around the 16th century mariners, merchants and so forth formed themselves into brotherhoods, fraternities or guilds, which were effectively religious charities for the mutual help and welfare of seamen and their dependants. Called Trinity Houses, three were set up in Scotland (Aberdeen, Dundee and Leith). England had Dover, Scarborough, Hull, Newcastle and especially – and possibly the oldest, dating perhaps from as early as the ninth century – Deptford, London's main port. Such companies comprised:

… godley disposed men who for the actual suppression of evil disposed persons bringing ships to destruction by the showing forth of false beacons do bind themselves together in the love of Lord Christ in the name of the Masters and Fellows of the Trinity Guild to succour from the dangers of the sea all who are beset upon the coasts of England to feed them unhungered and athirst, to bind up their wounds and to build and light proper beacons for the guidance of mariners.

Only London's league survived, to become known as 'The Master, Wardens and Assistants of the Guild, Fraternity or Brotherhood of the Most Glorious and Undividable Trinity and of St Clement in the Parish of Deptford Strond in the County of Kent.' In 1514 – mercifully – this was shortened to 'Trinity House'. That year Henry VIII had issued its first Royal Charter and appointed Sir Thomas Spert, 'Comptroller of the Navy' and master of the King's flagship, as Master of Trinity House, a post he held for 27 years. Thereafter the constitution was changed so that all elders or masters hold their position for only three years, either to stand down or face re-election by the Trinity Court of Brethren. This comprised the master, four wardens and eight assistants, a total of 13. This number had strong connotations with the Holy Trinity, literally 'three in one.' In 1604 James I raised this to 31 – still retaining the digits one and three – the additional 18 being referred to as elder brethren, a name deriving from the old brotherhoods. So-called younger brethren could be co-opted at the pleasure of the Trinity Court to vote during the election of the master or wardens. Sadly, in 1660, much of the early Trinity House archive was lost in the Great Fire of London, and then the current building at Tower Hill was bombed during the Second World War, so the wealth of material is not as complete as that for the NLB.

An official register of all seamarks in England and Wales was compiled and included visual cues such as tall trees, barns, farmhouses, spires towers and various other prominent landmarks that could not be removed without permission. Despite objections from numerous landowners, Queen Elizabeth I and her Privy Council ratified this Act of Parliament in 1566. It also recognised Trinity House as the official custodian of all British seamarks with the power to erect new marks and lights provided they were self-financing. Levies from shipping was still collected by the Lord High Admiral however, who also held the rights of ballastage. The dredging of sand and gravel to clear navigation channels, which could then be sold on as ships' ballast, caused further confusion. It was not until 1594 that the Queen commanded that ballastage and beaconage levies be surrendered to the Crown, which she subsequently awarded to Trinity House.

It had been the maritime aspirations of the Tudors that laid the foundations of the modern lighthouse service. But for a long time neither Queen Elizabeth I nor James I encouraged the building of lighthouses, fearing they would assist the enemy in times of war. In fact, only on three occasions have Trinity House extinguished its lights, firstly during the Dutch Wars (1665–7) but more British ships were sunk by this action than by the Dutch Navy! The other occasions were during the two World Wars nearly three centuries later. Although Trinity House did build a few lighthouses – its first one early in the 17th century at Lowestoft on the Norfolk coast – it functioned mainly by encouraging speculation and issuing charters from the Crown to landowners and entrepreneurs, enabling them to construct lighthouses and assume the levies.

In 1620 a private lighthouse was built at the Lizard despite open hostility from the local people who made considerable gains from plundering shipwrecks. Indeed a company of dragoons was needed to protect the builders. Soon afterwards other beacons were built, such as those at Orfordness, North Foreland and South Foreland. The success of some early lights was largely a hit and miss affair although the number of private ventures gained pace during the early 19th century, Trinity House was still reluctant to get involved since it would interfere with the livelihoods of pilots, and indeed of some of their own brethren. They did however accept responsibility for initiating the construction of Eddystone lighthouse in 1694. There was no particular strategy behind the erection of lighthouses in England and Wales, dependant more upon economic potential and speculation by willing landowners than by navigational necessity. Mariners soon began to complain about extortionate fees, so it was ultimately the greed of some landowners, their lack of standards and efficiency in maintaining lights – together with many obvious gaps in coverage – that finally forced Trinity House to react.

The Skerries, off Anglesey, North Wales

Early in the 19th century it was decided to bring them all under one unitary authority and Trinity House began a programme of assuming all these assorted private lights and standardising the collection of dues. An Act of 1836 accorded it compulsory purchase of the final 10 private lighthouses. The total compensation that the Corporation had to pay to the owners amounted to a staggering £1,182,546. The final negotiation concerned The Skerries off Anglesey, Wales. In Welsh these rocks are known as Ynysoedd y Moelrhoniaid, meaning 'the islands of the bald-headed seals'. They have attracted many wrecks in their time, but perhaps the most celebrated was the first Royal Yacht *Maria* on 25 May 1675. She came to rest on her side with her mast touching land, enabling just over half the passengers to scramble ashore. Over 30 souls perished. When a lighthouse came to be erected in 1714, the very first boat to approach with six workmen and the owner's son was lost. The original tower, topped by a coal-fired brazier, was completed three years later. In 1739 the lightkeeper and his wife, thinking themselves in splendid isolation, had a knock at their door ... to reveal a naked negro, the sole survivor of a ship wrecked on the rocks in the thick fog! The tower was later replaced and continued to provide a handsome profit for the owner until, understandably reluctant, he was finally bought out by Trinity House for an astonishing £444,984 – the final, most profitable and the most expensive light to be added to their books.

Although more remote, The Smalls share several features in common with The Skerries, and have an equally interesting history. The Smalls lie at the opposite extremity of Welsh waters, some 20 jagged rocks 32 km (20 miles) west of St David's Head in what used to be referred to as Pembrokeshire. Its lighthouse had been established in 1776, built to a unique design by a 26-year-old Liverpudlian violin-maker, Henry Whiteside. Somewhat loquaciously, a newspaper would later report:

> His undertaking was a sudden transition from the sweet and harmonious sounds of his own instruments, to the rough surging of the Atlantic wave, and the discordant howling of the maddened hurricane; and from the fastening of a delicately formed fiddle, to the fixing of giant oaken-pillars in a rock as hard as adamant!

He opted for an octagonal wooden tower, first assembled ashore, and then reconstructed on site. It sat on a frame of nine, massive oak timbers, each two feet in diameter, sunk into the parent rock. This arrangement allowed the sea to pass between, dissipating the force of the waves. Working against the worst that the elements could throw at them, his stalwart team – including some Welsh miners – completed the building in just over a year. But within months it was realised that this innovative structure had to be strengthened with many more oak piles.

The Smalls, off St David's Head, South Wales

Around 1800 one of the two lightkeepers on The Smalls died suddenly. His companion, with whom he was always arguing, feared that he might be accused of murder, so he made a crude coffin to hold the corpse and lashed it to the railing outside. Persistent bad weather opened the coffin to expose the dead body, and delayed the next relief for many weeks. Haunted by the apparition outside, waving its arms in the wind, the lone survivor was slowly driven mad. Thereafter, it is said, it became policy to station a third man on each isolated lighthouse.

Astonishingly one of The Smalls' keepers was to man this shuddering structure for 13 years. In 1833 a huge wave smashed up through the floor, it flattened the iron range and confined the petrified keepers within their leaking refuge for eight days before rescue. One of the keepers died of his injuries.

In 1836 Trinity House purchased the lease of The Smalls for £170,468. It remains the most isolated of all Trinity House stations. Whiteside's tower on stilts – albeit with constant and considerable repair – continued to function for over 80 years until James Douglass replaced it with the existing, conventional stone tower – 43 m high (141 ft) – in 1861. A helipad was added to the top of The Smalls tower in 1978. It was finally automated in 1987, and now has a ring of solar panels around the helideck.

After 1836 Trinity House was to embark on an ambitious programme of lighthouse building, but its golden age was undoubtedly between 1879 and 1900 – when it was to construct or modernise at least 50 stations around the English and Welsh coasts, with new rock stations built at Wolf Rock, Longships, Bishop Rock and Beachy Head. During this time Trinity House utilised some distinguished engineers, with of course John Smeaton at the forefront.

James Walker (1781–1862) was born in Glasgow but articled as an engineer to his uncle in London, where he became involved in dock and harbour buildings. When he was appointed consultant engineer to Trinity House he or his firm became responsible for 29 towers. Where there was room – such as on South Bishop (1839) and The Skerries (1846) – he incorporated some stylish ancillary buildings and courtyards. He provided The Smalls (1861) and Wolf Rock (1870) with the first proper water-closets, but more importantly he

South Bishop

Wolf Rock, off Land's End – Longships lighthouse beyond

employed a stepped base to break the waves at these exposed sites. At Eddystone and Bell Rock particularly rough seas often swept up the tower and even obscured the light. Despite his unsatisfactory efforts with the first Bishop Rock in 1847–50, he greatly improved his designs for rock towers. He was even to incorporate perpendicular sides at The Needles in 1859, completely eschewing Smeaton's tree trunk model at Eddystone.

Walker had designed the first ill-fated iron tower on Bishop Rock in the Isles of Scilly, working with Trinity House's constructional engineer Nicholas Douglass, with whom he then co-operated on its replacement stone tower year or two later. Already Douglass was grooming his two sons as engineers, and both worked alongside him on the first two Bishop Rock lighthouses. With Walker's designs, James Nicholas Douglass (1826–98) was resident engineer at The Smalls from 1855–60, and Wolf Rock in 1870. After his appointment to Trinity House's engineer-in-chief he designed 20 new towers, including the new Eddystone, and the reconstructed Bishop Rock, for which he was knighted. Sir James's sons William and Alfred both went on to become lighthouse engineers, and with his father and brother

Skellig Michael, County Kerry, Eire

constitute the nearest English equivalent to the dynasty already established in Scotland by Robert Stevenson.

After Wolf Rock and 26 years with Trinity House, Sir James's younger brother William (1831–1923) went on to spend the next 20 years in Ireland as engineer-in-chief to the CIL, for whom he designed eight impressive new towers, including the second Fastnet. The first Fastnet tower had been constructed by George Halpin in 1854 but ultimately he was not to build as many lighthouses as his father, also George, whose first effort was Inishtrahull in 1812. George Senior's first rock tower was Tushkar in 1815, followed by both lights on the Skellig Michael and then Eagle Island, amongst many others, some with assistance from his son. But not much is known about the Halpins, despite their having been responsible for over 50 lighthouses around the Irish coast.

Nowadays these tiny isles of ours have one of the most well lit coasts anywhere. In Ireland, CIL are responsible for some 80 lighthouses and numerous other beacons, buoys and unlit perches. Trinity House is responsible for only 65 lighthouses but 10 light vessels/light floats, 412 buoys, 19 beacons, 48 radar beacons and seven satellite navigation reference stations. On the other hand, although the NLB is a much smaller institution than Trinity House, it operates no fewer than 205 lighthouses along with 164 buoys, 26 beacons, 35 AIS stations, 29 racons and four satellite navigation reference stations. About 25 of the major lighthouses are built around the Scottish mainland with Kinnaird Head and Cromarty for instance close to town centres. Others like Cape Wrath are remote, yet most mainland lights are still reasonably accessible, at least compared with rock stations such as Skerryvore or Muckle Flugga. The rest are located on islands, some extensive with a permanent population, others are on uninhabited islands, a few on isolated on rocks barely exposed by the tides.

Geographical location determined how lighthouses were classified and, until automation, how they would be manned. 'Shore stations' allowed the keepers to have their families accommodated at or close by the lighthouse; they worked in shifts to a five-day week, with normal annual holidays. 'Island family stations' differed only in that they were located on an island rather than the mainland. A few such stations were built as or downgraded to one-man operations, the light requiring no night watches so the role was almost one of caretaker. 'Relieving stations' were located on the mainland or large, inhabited islands but sufficiently remote and difficult of access for the families to be accommodated elsewhere. 'Rock' lighthouses were on small uninhabited islands or barren rocks, reefs or skerries offshore. Both these situations required a crew of six men, working to a fixed routine of 28 days on and 28 days off, three always manning the lighthouse whilst the others were ashore with their families. They commuted back and forth to their work by boat or, latterly by helicopter. Rock lighthouses were considered the 'front line' of the profession.

The story of our lighthouses is not only influenced by geography but is also affected by social and economic history. From the Middle Ages Scotland had maintained a lively trade in salt fish, wool, hides, and later, wheat, rye and malt, with several North Sea ports, in exchange for fine cloth, flax and good wines. Leith, Dundee and Aberdeen were major centres of such commerce. Various seamarks stood at harbour entrances for the benefit of such merchant vessels and also the local fishing fleets. In 1563 two beacons had been established at the mouth of Leith harbour, one to be later damaged by an English ship. Three years later a seamark was known to have been in position at Aberdeen. Fifty years on, coal mining, salt production, fishing and the North Sea trade demanded further marks or beacons on 'all the craiges and blind rocks' within the Firth of Forth 'for the crydit of the country and saftie of strangers.' But these formed a confusing assemblage against the coal fires from the Fife saltpans and limekilns. By now merchants were actively petitioning the Crown for a lighthouse on the Isle of May.

The Isle of May is a horizontal volcanic flow formed some 300 million years ago during Carboniferous times when giant trees fringing steamy, tropical lagoons would, much later, form the extensive coal deposits of Scotland's central belt. It was a time of great volcanic ac-

The Isle of May, Firth of Forth, Scotland

tivity, spewing flows of hot lava from vents in both Fife and Lothian. Hard plugs of residual lava resisted erosion to form the Bass Rock and Edinburgh Castle Rock, while horizontal lava outflows or sills from long-gone volcanoes can still be seen at Arthur's Seat, for instance. The May – another such sill – lies 8 km (5 miles) offshore just where the Firth of Forth turns into the North Sea.

At 45 ha, May is the largest of several islands in the Firth, mostly comprising low, flat rocks, with its southwest cliffs rising to only 46 m (150 ft). Although Neolithic pottery has been found, the island's first recorded inhabitant seems to have been St Ethernan (later known as St Aidan) on a mission to convert the people of Fife to Christianity in the ninth century. He was killed by the Vikings in 875 and buried on the Isle of May where a chapel was later erected in his name. His martyrdom led to the island becoming an important place of pilgrimage. It is said that 10th century monks had kept a beacon burning there to guide voyagers on the pilgrims' way past the May to the holy shrines at St Andrews.

By 1549 its religious significance had waned and the Isle of May passed into private hands, latterly those of Alexander Cunnynghame who then applied to the government to install a light on his island. Five years later, in 1636, the beacon was finally erected, to become Scotland's very first permanent lighthouse.

At a height of 12 m (40 ft), the ornate, three-storied tower was 7.6m (25 ft) square, had an internal spiral staircase and a flat roof, surrounded by a castellated parapet. The lighthouse keeper and his family lived upstairs. If the keeper had to venture out fishing at night to sustain his family, no light was shown, so one or two assistants were also appointed.

Scotland's first lighthouse on the Isle of May

A large metal grate was mounted on the roof for the coal-fired beacon. This consumed the best part of a ton of coal each night, up to three on stormy nights – some 200 t every year! Hot ash was tipped over the parapet wall.

All ships sailing from or arriving at ports between Dunnottar Castle to St Abb's Head, were to pay 2 shillings Scots per ton (2 pence sterling) if they belonged to the kingdom of Scotland, and four shillings (4 pence) for foreigners (which for many years after the 1707 Union of Parliaments still included English ships!) The levies were collected at the Custom Houses in the Fife ports for the proprietors of the island who, in turn, paid a commission to the Scottish crown. The dues were later revised down several times but, in 1800, it is estimated that 450,000 t of shipping were being taxed.

In those days coal was cheap. In 1799 for instance, 366 t cost only £142.11s, with an additional freight charge to the Isle of May of £83.10s. On the other hand, the keepers' total wage bill for the year only amounted to £70. They had to manhandle the coal, not only up on to the roof, but also from the pier when it first arrived on the island. It was not until 1786 that a horse was brought in to assist with the latter task, and a rope and pulley system was installed on the tower's roof. The wall of one corner can still be seen to have its edge rounded off by the constant hammering of sacks or pans of coal being raised to the beacon on top. The grate was also enlarged to improve the beacon.

According to D Alan Stevenson, the original circular grate had been 85 cm (33 ½ in). across the top, 71 cm (28 in.) broad at the bottom and 53 cm (21 in.) deep while the enlarged one was 1 m (3 ft 6 in.) in diameter and 60 cm (2 ft) deep. In December 1800 this was said to require 4 cwt of coal every hour, doubling the annual consumption. It was however deemed the most powerful coal light in the kingdom. Notwithstanding, the beacon was still largely ineffective in bad weather when it was most needed. Although a glass screen was erected around the fire 'it smoked the glass so much as to render the light useless.' Even when the beacon was operational there was still the risk of confusion from industrial fires on the coast.

In December 1810, during the Napoleonic Wars, two Royal Naval frigates, *Nymph* and *Pallas*, were wrecked near Dunbar due, apparently, to the fire of a limekiln on the East Lothian coast being mistaken for the May light. Although both ships were lost, only nine men were drowned out of joint crews of some six hundred. The Admiralty petitioned that the Commissioners of Northern Lighthouses (established in 1786) take over the Isle of May. At first the then proprietor, the Duke of Portland, resisted until in 1814 he eventually settled for a selling price of £60,000 – £3,000 less than he had originally demanded.

Two years later, as we will presently discover, the light was modernised, replaced by an oil lamp and reflector arrangement, housed in a handsome and substantial tower designed by Robert Stevenson. The old beacon blazed for the last time on 31 January 1816, after a service of nearly two centuries. Stevenson had proposed demolishing the original lighthouse but Sir Walter Scott pleaded with him only to ruin it 'a lá picturesque,' and so it remains to this day. For a time it served as a lookout for smugglers, for six bored marines 'on a sloping wooden bench … to recline upon, ruminate, snooze and smoke tobacco.' According to one Reverend Gordon in 1865, 'they are rather chatty and come to the door with their spy-glass, to enlarge the prospect, for the delight of any visitor who vouchsafes them a quid of Virginian [tobacco].' The building later gave accommodation for Forth pilots and fishermen. Reduced to a single whitewashed storey, in 1958 it became a listed building and Ancient Monument of considerable historic significance.

Hitherto, Scottish trade was mostly with Scandinavia and the Low Countries through the Firth of Forth ports while, in 1687, two leading lights had come to be exhibited from Buddon Ness, at the mouth of the Firth of Tay near Dundee. A guiding light had been erected at Portpatrick on the Solway around 1680 but the Act of Union 1707 opened up Atlantic trade routes into Scotland and, with Dumfries an important port at the time, a light was built at Southerness on the Solway in 1749. By 1760 however there were still only six lighthouses in Scotland.

The Firth of Clyde was determined to grab a share of trade, although Greenock was the upper limit for deep-water navigation. Thus Port Glasgow came to be created a few miles further upriver, as a rival centre with better Glasgow connections. In 1756, an Act of Parliament established the Cumray Lighthouse Trust – the very first statutory lighthouse authority. Composed of the Provost and Magistrates of Glasgow, the bailies of Greenock and Port Glasgow and Clyde landowners with a specific stake, the Trust's first priority – and indeed their whole raison d'être – was the establishment of a lighthouse on Little Cumbrae, opposite the fishing village of Millport on Great Cumbrae.

Little Cumbrae, Firth of Clyde, Scotland

Barely a couple of miles long and a mile wide some two acres of ground was feued from the owner for £2 per annum so that, on the highest point of Little Cumbrae 60 m (400 ft) above sea level, a circular 8.5 m (28 ft) high stone tower could be constructed within the year. It supported an iron cage or grate in which the coal-fired beacon was ignited nightly. In 1757 the first two lighthouse keepers (Peter Montgomerie and Thomas Fairlie) were appointed, only to be replaced before the following Whitsun by David Pollock. When he finally died in 1804 his wife and son took over, still on the same wage of £30 per annum forty years later!

Maintaining Port Glasgow and any further access upriver necessitated considerable dredging operations and buoyage as well as lighthouses. First John Smeaton (the marine engineer who had built the fourth Eddystone light) then John Rennie, Thomas Telford, James Watt and other great civil engineers of the time, became involved at various stages of these developments. This time the dues would be paid not to private individuals, as at the Isle of May, or to a fraternity of sailors as at Dundee, but to Trustees who would use them exclusively for the upkeep of the light. Ships that 'passed or repassed the said lighthouse' were charged accordingly, one penny per ton burden for British vessels on foreign voyages, two pence for foreign ships, a half-penny for 'home traders' up to 30 tons, but, strangely, two pence per ton for home traders between 15 and 30 tons. Thus, for the first time, an authority was set up for the principal provision of lighthouses, and they went on to install other lighthouses at Cloch in 1796, and then Toward in 1812. But by then it was obvious that the lofty position of the Cumbrae light was too often obscured in mist and fog, so a lower light had to be built on the shore. The firm of Thomas Smith and Robert Stevenson were involved in this in a private capacity although much of their lighthouse work was for the state-controlled Commissioners of Northern Lights who built and maintained the other lights downstream and beyond. The Clyde Lighthouse Trust replaced the Cumray Trust in 1871 and it is now the Clyde Ports Authority, which continues to be responsible for their three lighthouses – Cumbrae, Cloch and Toward.

It is indeed remarkable though that the labour-intensive and often ineffective coal-fired operation persisted for so long. And, it was not without its human tragedies either. Legend maintains that returning from the Isle of May after that lighthouse had been erected, Cunnynghame – its instigator – was drowned, in a storm whipped up by an old woman from Crail, Eppie Lang, because he had wronged a local girl. Some years later Eppie Lang let slip that she had enlisted help from the devil, so she was burned as a witch.

A more ancient legend from the Isle of May concerns Thenaw, the daughter of a pagan king of the Lothians. Refusing an arranged marriage, surviving the brutal punishment from her father and pregnant to boot, she was condemned as a witch. She was cast adrift in a coracle from Aberlady and abandoned to her fate. Her first landfall was on the Isle of May until finally a favourable wind fetched her up near Culross in Fife. In the care of St Serf and his monks, she gave birth to a son – Kentigern or Mungo, later to be the patron saint of Glasgow. Out of pity for her suffering however, the abundant fishes of Aberlady Bay had followed her coracle and only took their leave off the Isle of May, where they have remained ever since. *The Life of St Kentigern* added how 'there was such an abundance of fish that many

Old print of Cloch lighthouse, Firth of Clyde

fishermen from all the shores of England, Scotland, Belgium and France come because of the rich fishing, and are welcomed into the harbours of the Isle of May.'

There have been countless wrecks and naval battles on and around the Isle of May in its time, but the most tragic of all incidents, without any doubt a true story, took place at the old coal-fired beacon itself – designed to save lives, not to take them. One night in January 1791, lightkeeper George Anderson, his wife and four children were overcome with fumes from the ash pile just outside. Only a baby survived, Lucy, who was subsequently brought up in Fife. Almost lifted from the realms of fiction, at 16 years old Lucy married one of her rescuers, Henry Downie, 22 years her senior. They emigrated to North America.

We will return to this tragic tale at the end of the book.

Lucy Anderson in later life

five

A Most Dangerous Situation

I will venture to say there is not a more dangerous situation upon the whole coasts of the Kingdom, or none that calls more loudly to be done than the Cape or Bell Rock.

Robert Stevenson

Lighthouses are familiar to us all. And the Scottish coast is particularly rich in them. Although all are individually suited to their particular location, they possess a certain leitmotif and pleasing theme to their design because – what is less well known perhaps – one family, over several generations, built most major Scottish lighthouses . The Stevensons. This dynasty went on to influence lighthouse design to this day, and also advised on the construction of lighthouses worldwide. In addition, their family firm were responsible for countless beacons, harbours and bridges. It is indeed odd that it was the black sheep, or – as one commentator would have it – 'the wreckling of the family', who has become the most famous … the writer Robert Louis Stevenson.

Several eminent books have already been written about the Stevensons and Scottish lighthouses, notably *A Star for Seamen* by Craig Mair (1978), *The Lighthouse Stevensons* by Bella Bathurst (1999) and *Dynasty of Engineers* by Roland Paxton (2011). The present volume however, while it does celebrate the Stevensons, is intended more to look at the natural history associated with lighthouses and lightkeepers, mainly in Scotland. Although no longer up to date, I found RW Munro's *Scottish Lighthouses* (1979) and Keith Allardyce and Evelyn Hood's *At Scotland's Edge* (1986) both exceedingly relevant and readable. But for an encyclopædic review of Scottish lighthouse development, design and mechanics, up to the present, look no further than Alison Morrison-Low's monumental *Northern Lights* (2010). None of these however tell much about the role lighthouse engineers and keepers – encouraged by naturalists – contributed to science, and to natural history in particular. That is the fascinating story told in this volume, woven around the history of lighthouses and keepers, and of the remarkable Scottish dynasty of engineers.

The Stevensons were based in Edinburgh, an environment heavily influenced by what is known as the Scottish Enlightenment. Thus the city was particularly conducive to encouraging not just their engineering skills but also their diverse interests and hobbies. Amongst these were the natural sciences – geology, meteorology and natural history. In their long and proud history, it has transpired that lighthouses were often built in wild places, rich in wildlife. Thus, besides a degree of boredom through isolation, it is not surprising that many lightkeepers came to occupy their spare time observing the environment around them.

Towards the end of the 18th century fishing, trade and travel had greatly increased shipping in Scottish waters. All vessels trading between the Clyde and Baltic ports had to sail round the north of Scotland. A Dundee merchant, lawyer and Member of Parliament, George Dempster (1732–1818), actively promoted the provision of lighthouses around the Scottish coast. As a result a Commons Committee was set up, drafted by a Writer to the Signet called John Gray, which recommended establishing a board of trustees or commissioners to pursue the matter. An unprecedented succession of storms in 1782 highlighted the urgency.

With its roots prior to the Union of the Crowns in 1603, Trinity House had no jurisdiction over Scotland, so an Act of Parliament was quickly passed in 1786 to establish the Commissioners of Northern Lighthouses (later the Northern Lighthouse Board, or NLB) in Edinburgh, and indeed the Commissioners of Irish Lights (CIL) in Dublin. These were probably the first authorities in the world with the sole duty of managing lighthouses on a national scale. In the spirit of Parliamentary Union in 1707, and with the government backlash against all things Jacobite (or indeed Scottish) after the 1746 Jacobite Rebellion, for a long time afterwards Scotland would be referred to as 'Northern Britain'. In 1815 the Isle of Man recognised their superior expertise and opted to become affiliated to the NLB rather than Trinity House; so perhaps the inclusion of the word Scottish might have become inappropriate anyway. John Gray – of whom it was said was 'famous for drinking punch, holding his tongue and doing his job quietly' – became the NLB's first secretary until retiring in 1811.

The commissioners were not selected for their maritime experience but were made up of a curious panel consisting of two Crown Officers for Scotland, the Lord Advocate and Solicitor-General, the Lord Provosts of Edinburgh and Glasgow, the Provosts of Aberdeen, Inverness and Campbeltown, and the Sheriffs of certain maritime counties. Nonetheless, in order 'to conduce greatly to the security of navigation' they quickly authorised the building of four new lighthouses at strategic headlands around the north of Scotland – Kinnaird Head in Buchan at the easternmost extremity of the Scottish mainland, North Ronaldsay at the northern tip of Orkney, Scalpay on Harris in the Minch, and the Mull of Kintyre between Scotland and Ireland.

The 1786 Act of Parliament gave the Trustees of the Northern Lighthouses the power to purchase land, levy dues and borrow funds. At first they were only permitted to raise £1,200, wholly inadequate for the task in hand. They set the dues for shipping at one penny per ton on British ships, and two pence (2d) per ton for foreign vessels, with exemption on whalers and naval ships and vessels trading in summer with the Russian ports of Archangel and the White Sea. These would only come into effect as soon as the first four lighthouses were completed. So, undaunted, they set to the task in hand.

Initially, the NLB had difficulty locating an engineer but eventually had a response from an English amateur Ezekiel Walker (1741–1834), who had some experience building lighthouses. He agreed to build only one lighthouse himself but, for the sum of 50 guineas, would give instruction to someone on the other three. Already in receipt of *Observations on Lighthouses by Reflectors, with a small model* for the first meeting of the commissioners, the man they would ultimately appoint as engineer was its author Thomas Smith (1752–1814), an 'ingenious and modest … tin plate worker.' He was despatched forthwith to Norfolk for instruction by Walker and, after only a couple of months, was in Fraserburgh ready to make a start on the Kinnaird Head lighthouse.

Thomas, the only child of a skipper from Broughty Ferry who drowned at sea, was born in Dundee. Indeed his father had been a member of the Fraternity of Seamen who had built the Buddon lights. His mother was the daughter of a Leith shipmaster so, like his father and grandfather, he too longed to go to sea. But he became apprenticed to a Dundee tinsmith at a time when the city enjoyed a thriving whaling industry. Wealthy householders were quickly replacing their domestic tallow candles with oil lamps. Around 1770 Smith moved to Edinburgh to live with his mother's family and set up a successful business manufacturing assorted metal appliances and oil lamps.

Thomas Smith quickly adopted recent innovations by Aimé Argand in Geneva who devised a circular wick surmounted by a glass chimney, which proved more economical in

Modern street lighting in Edinburgh

oil, while enabling the draught to be controlled so as to burn more brightly. Smith enhanced the efficiency further by substituting polished tin reflectors with parabolic ones carefully inset in plaster with 350 facets of mirrored glass. Ezekiel Walker claimed that it was he who had devised the original reflectors (which might explain Smith's unduly short apprenticeship under him, and why he seems never to have built a Scottish lighthouse), but Smith certainly improved upon the design. Such lamps he used to provide street-lighting first in the Old Town, then in 1787 in the New Town of Edinburgh and – with his detailed submission mentioned above – had convinced the Commissioners that these could be installed in lighthouses.

Smith owned an Arctic whaler so it is not surprising that he promoted the use of whale (spermaceti) oil in his lamps. This was to remain the preferred illuminant in most Scottish lighthouses for 50 years. (It produced a lot of smoke however, hence the need for a flue above the lamp to dissipate fumes. The adoption of colza (rapeseed oil) reduced this problem). Smith's own son James (1783–1820) and grandson Thomas eventually carried on the street lighting side of the business while Thomas senior himself, and then his stepson Robert, concentrated on lighthouses and marine engineering. At first Smith undertook his work for the Commissioners of Northern Lighthouses for free until in 1793 they awarded him with a salary of £60 per annum, with expenses.

On the small clifftop at Kinnaird Head, the lantern was placed on the roof of a redundant (but nonetheless expensively purchased) 16th century castle, a task supervised by Smith

Kinnaird Head, Fraserburgh. Print by William Daniell c.1816

Kinnaird Head, Fraserburgh, now the Scottish Lighthouse Museum

himself to the design of an Edinburgh architect, Alexander Kay. It was the most powerful light of its time, with 17 reflectors arranged in three tiers. After it was lit on 1 December 1787, one Stornoway skipper commented: 'If the light is kept in as good a state as when I saw it, I think it will be superior to any lights on the coast, and equal to any I ever saw.' It is appropriate that this first NLB undertaking in its majestic setting – after it was superseded by a small automatic light nearby – should now be the location of the highly acclaimed Scottish Lighthouse Museum: well worth a visit.

The Mull of Kintyre presented more of a challenge to Thomas Smith, all the materials having to be hauled with packhorses around 19 km (12 miles) overland to the site, 90 m (295 ft) above the sea. A Provost of Campbeltown judged: 'No stranger could be supposed to contract for a building in a place so difficult of access and so remote unless at a very extravagant sum.'

The station was rebuilt in the early 1800s and in 1906 the fixed light was altered to a more powerful, group-flashing light. It was one of the Mull keepers who first arrived on the scene on 2 June 1994 when a military Chinook helicopter, on its way back from Northern Ireland, crashed nearby; all 29 passengers died. The Mull of Kintyre was automated two years later.

Mull of Kintyre lighthouse, Argyll

Eilean Glas on Scalpay, Harris, was the first lighthouse to be built in the Outer Hebrides (although monks had kept a beacon alight on one of the Monach Isles; later in the 15th century, it is said that a torch beacon had been maintained on Scalpay itself). George Shiells and two other masons were assigned to Eilean Glas but bad weather halted their work so they were sent to work on the Kintyre station in the meantime. Smith took particular care to ensure that the keepers' accommodation on Kintyre was secure enough, so the light was not finally exhibited until October 1788. North Ronaldsay, at sea level, necessitated a 21 m (70 ft) tower whose construction was undertaken by Orcadians John White and James Sinclair; it was lit in October 1789.

Meanwhile, Thomas Smith was told that a man called Campbell, tenant to the proprietor Captain Macleod of Harris, had already begun at Scalpay. Earlier Campbell had submitted to the NLB an estimate of £500 for the work, to be done by local men for only one shilling a day, but it had been rejected. Smith preferred his own Edinburgh masons. So, probably in a fit of pique, Campbell had taken it upon himself to begin the work, equipping his men with more tools and ropes than necessary, many of which proved unfit for the purpose. He had also constructed a quay, which was soon broken up by the sea so never used. But more worrying, he was building the tower four feet wider at its base than Alexander Kay's original design.

After a hasty and difficult voyage to Scalpay, and to avoid unpleasantness between Campbell and the future lightkeepers there, Thomas Smith turned diplomat and compromised, even compensating Campbell a small sum for his trouble. George Shiells resumed the work. Although the lightroom was raised 7.5 m (25 ft) above the ground to bring it to about 22 m (73 ft) above sea level, wave action demanded the deployment of

Eilean Glas lighthouse, Scalpay, Isle of Harris. Print by William Daniell c.1816

thicker glass panes. In May, Thomas Smith delivered the panes himself all the way from Edinburgh, aboard a Wick vessel called *Kelly and Nelly*. The first keeper Alexander Reid and his family were aboard with him. And so the light was eventually exhibited, along with that on Ronaldsay, in October 1789.

Smith had found that each of the four lights had presented their own special circumstances, dependent upon their situation. Thus his standard light with 48 reflectors had to be modified to suit; Kinnaird and Kintyre had 36 reflectors, North Ronaldsay 31 and Eilean Glas only 28. In future every lighthouse would require its own individually tailored light unit. Nonetheless four effective and impressive lights had now been established – within a period of only three years. Robert Stevenson was later to admit how these early lights were built: '… on the smallest, plainest and most simple plan that could be devised, and with such materials as could be easily transported and most speedily erected.' The lightrooms were of timber with cast-iron window sashes and copper sheeting on the roof outside and fireproof metal sheeting on the ceilings and floors. One or two would be upgraded or superseded later by his stepson Robert Stevenson.

Once built these lighthouses had to be manned. Robert Louis Stevenson recorded how: 'A whole service, with its routine and hierarchy, had to be called out of nothing; and a new trade, that of lightkeeper, to be taught, recruited and organised.' The first NLB lightkeeper, appointed in 1787 on a wage of £30 per annum with a garden, pasture for a cow and enough fuel for domestic use, was James Park at Kinnaird Head, a former ship's master: he was also expected to find an assistant to be with him every night. Every morning he had to clean the reflectors and lantern panes and top up the lamps with whale oil, so that they could be lit half-an-hour after sunset, then attend them throughout the night until half-an-hour before sunrise. He faithfully maintained this routine for nearly 10 years until ill health forced him to retire at 80. On a sea voyage shortly afterwards, Park was captured by the French and imprisoned for some months. On his release he and his wife were allocated a small room back in Kinnaird Castle.

Park's successor at Kinnaird was George Gray, another retired mariner, who was to use his family as his assistants, which became not an uncommon arrangement in the service. Robert Stevenson, Smith's apprentice and stepson, would come to disapprove of this on the grounds of encouraging sloppiness and ill-discipline, with no likelihood of any failure of duty ever being reported. So, at every station, Stevenson introduced the post of assistant keeper. Accepting that – living and working so closely together – they might not necessarily get on with one another, one keeper could be better relied upon to report back on the other's deficiencies. George Gray received Kinnaird's first assistant lightkeeper, James Lindsay, in the 1820s. Sometime later – there is no date, nor are the parties named – Stevenson was to record a quarrel at Kinnaird between the two keepers' families regarding a daughter's 'misfortune'! Gray's successor in 1827 was Peter Ewing, already a serving keeper from another station, who in turn would be succeeded by John Reid in 1843. Three years later and again four years after that, Alan Stevenson – Robert's eldest son and heir and obviously inheriting his father's high standards – would fine Reid and his assistant 30 shillings each, and then a further guinea, for keeping a dirty lightroom. At the time a principal keeper's wage was about £55 and that of an assistant £45.

Thomas Smith's choice for Kintyre's first lightkeeper was Matthew Harvie, a local man, whose family had for some years before kept a light burning in the kitchen window of their croft to warn seamen off the rocks. He was the first of what would become a family line of lightkeepers. But at North Ronaldsay another former shipmaster had to be dismissed after embezzling the stores. At Scalpay, in complete contrast, the first lightkeeper Alexander Reid was ultimately pensioned off after no less than 34 years' service. In 1816 the artist William Daniell wrote of him:

Few abodes, it may be imagined, can be more lonely and dispiriting, yet the man of constant duty here … a fine old weather-beaten mariner, [was] so fitted to his station, either by natural temperament or by long habit, as to fulfil his duties not only without repining, but even cheerfully … he had never been cheated of an hour's sleep by the stirrings of ambition, the thirst of lucre, or the corrosion of care.

Meantime, developments continued in the Firth of Clyde, where the Trustees for Clyde Lights had already established several independent lights. Merchants continued dredging

Pladda lighthouse, Firth of Clyde

and realigning the river, enabling ships to sail right into the heart of Glasgow, which was destined to become the second greatest city of the British Empire. The next NLB lighthouse recommended in the 1786 Act was to be on the island of Pladda, off the southeast corner of Arran. George Shiells, the mason used at Kintyre and Scalpay, was soon at work and it was lit in 1790, but its beam had to be distinguished from the fixed lights at the Mull of Kintyre, on Cumbrae and at Copeland on the Irish coast. So a lower light 6 m (20 ft) below the main one was added at Pladda, a system that was to operate there for 100 years.

On the Clyde it had become evident that the hilltop light on Cumbrae was often obscured by cloud or mist and, since the Pladda light was considered by local mariners to be 'the best light they of anywhere', all agreed that Cumbrae would be improved by conversion from coal to oil lamps. Smith sent his assistant Robert Stevenson to supervise construction of a new lighthouse on the shore, which was lit in October 1793.

The previous year Stevenson had cut his teeth in the business with the installation of a harbour light at Portpatrick in Galloway. The Clyde authority now felt that another lighthouse was needed on the Clyde, at Cloch Point. Thomas Smith was commissioned to add the lantern, with young Stevenson's help, and which was lit in August 1797. By now Smith had also converted Buddon Ness from coal to oil.

There was also a plan to open up the Pentland Firth as a shorter shipping route than round the north of Orkney; but it was a notorious passage – known as 'Hell's Mouth' – which Robert Stevenson recognised as 'well known for its tempestuous seas, and cross-running

Stevenson's replacement for Little Cumbrae lighthouse

Cloch lighthouse as it is today (compare with p72)

tides.' Another double light was erected on the Pentland Skerries in 1794; two towers 24 m (80 ft) and 18 m (60 ft) high and 60 ft apart. Built by Orkney masons the work was supervised by Robert Stevenson, his first official task for the NLB. Two other workmen were Johnny and William Gray, South Ronaldsay born and bred. Johnny was later taken on for the building of the Bell Rock, where one of the rock features exposed at low tide – Johnny Gray's Landing – is named after him. William became the first Pentland Skerries boatman, based in his native island and began a family tradition that spanned five generations or 134 years, ending with William Budge in 1955. In 1814, on an inspection cruise with Robert Stevenson, William Gray found the Pentland Skerries keeper 'an old man-o'-war's man, of whom Mr Stevenson observed that he was a great swearer when he first came, but after a year or two's residence in this solitary abode, became a changed man.'

Pentland Skerries lighthouse, Orkney

Light dues had brought the newly fledged NLB an income of £1,477 in 1790 but by 1802 this had risen to £4,386, encouraging them to consider building new lights. The island and rocks of Inchkeith in the Firth of Forth had always presented a hazard to ships entering the naval port of Leith so Stevenson identified a suitable site for a lighthouse in an old fort dating from 1564 and Mary Queen of Scots. Inchkeith came to be exhibited in September 1804, with Stevenson incorporating reflectors of silver-plated copper instead of his mentor's old mirror-glass; a plaque on the handsome tower and accommodation block credits Thomas Smith as being the engineer. Notwithstanding, Robert was now effectively running the lighthouse side of the family business, having been made a full partner in 1800.

An unlit tower of masonry was built as a seamark on Sanday, Orkney in 1802. When a new light was proposed at Start Point on the island it was at first resented by the penurious community as it might interfere with the plunder they gleaned from all the wrecks in

Sanday lighthouse, Orkney. Print by William Daniell c.1816

the vicinity. No fewer than 22 had occurred in the previous 14 years. For a time this new tower came to replace the light on North Ronaldsay altogether. A stone ball from the original tower on Start Point (now classified as an historic monument) was placed atop the redundant structure on North Ronaldsay (and it was not until 1854 that the present tall and elegant, brick-built tower came to be built alongside – at 42 m (138 ft) and with 176 steps the tallest land-based lighthouse in the UK). In 1806 Start Point (with its unique and distinctive black vertical stripe added in 1915) exhibited Scotland's first revolving light. In truth Stevenson had wanted to employ this innovation on Pentland Skerries some years earlier, being a cheaper method of distinguishing different lighthouse beams than constructing the two separate towers. On a tour of English lights in 1801 he had been impressed by St Agnes in the Scillies and Cromer where the reflector frame revolved by clockwork 'exhibiting a brilliant light once in every minute'. Pentland Skerries was to be strafed by a German plane in the Second World War – only minutes after the lightkeeper had left the lamp room.

Already, lighthouse operations in Scotland had become so widespread that Thomas Smith, as NLB Engineer, had to undertake annual inspections by sea. At this time Britain was at war with France and Smith came close to being captured by a French squadron. Press gangs from the Royal Navy were another hazard, in addition to the accepted risks negotiating dangerous waters and storms. All this was to impose gruelling physical demands on Smith

North Ronaldsay, old beacon tower and current lighthouse

so, from 1797, the young Stevenson became responsible for all the NLB inspection voyages, though yet still not without risk. On 9 October 1794, after completing his work at Pentland Skerries, Stevenson sailed south in a Stromness sloop *Elizabeth* but, becoming becalmed off Kinnaird Head and with urgent business at home, Stevenson was rowed ashore to continue overland to Edinburgh. Only hours later on its return north, a gale blew up and the *Elizabeth* was lost with all hands. A narrow escape for Robert Stevenson – and indeed for the future of Scottish lighthouses!

Thomas Smith had been married twice before but his first wife had died, along with three of their five children. His new wife and their only child also died. With his business expanding and demanding more time away from home, his kindly neighbour Jean (sometimes called Jane) Stevenson née Lillie (1751–1820), who had been a friend to both his wives, offered to help look after his children. They married in 1787. Smith took on Jean's 19-year-old son, Robert Stevenson (1772–1850), as his apprentice.

Robert's father Alan had worked as a merchant but died of fever in the West Indies when his son was only two. Apparently Jean then married again but James Hogg deserted her, and nothing is known about their two children. Robert grew up with Thomas Smith's two surviving children and was to marry his daughter Jane (1779–1846) in 1799: she was 20 and he 27. Robert was made a business partner in his stepfather's business the following year. Thomas Smith finally relinquished his post as NLB Engineer to his stepson in 1808. Smith died in 1815 having constructed 10 lighthouses, which founded an international reputation for the NLB.

Attending university classes in engineering, natural philosophy, mathematics, chemistry and natural history Robert never actually graduated. But apprenticed to a lamp maker and tinsmith, he went on greatly to enhance the reputation of the family firm and of Scottish lighthouse engineering. In 1814 his friend Sir Walter Scott was moved to comment: '… a most gentlemanlike, modest man, and well-known by his scientific skill.' Stevenson and Scott had first met in 1795 through the Edinburgh Volunteers (the contemporary equivalent of the Home Guard in the Second World War) formed a year earlier, and in which Stevenson's stepfather Thomas Smith had been a captain.

Robert's eldest son Alan would later remember his father thus: 'He was a man of sincere and unobtrusive piety and had perseverance, fortitude and self-denial, and an enthusiastic devotion to his calling. Add to this a deep quiet sense of humour and great physical strength

… He hated parties, except in his own house, had a fiery temper, easy to arouse.' Later grandson Robert Louis added: 'He had moved since boyhood as a pioneer among secluded and barbarous populations: his knowledge of the islands and their inhabitants was perhaps unrivalled.'

Robert and Jane were to have 13 children but – almost like a curse on the family – only five survived – Jane, Alan, Robert, David and Thomas. Robert Louis, lamented how: 'Never was such a massacre of the innocents; teething and chin-cough and scarlet fever and smallpox ran the round; the little Lillies, and Smiths and Stevensons fell like moths about a candle.'

Alan, David and Thomas all followed in their father's footsteps – not always willingly at first. Then David's two sons – David Alan (David A) and Charles – became engineers, as did Charles's son David Alan (D Alan).

Robert frequently wrote at length to his sons about natural curiosities that he had encountered on his travels, arousing their interest in all around them. As well as his engineering affiliations Robert was also an elder of the Church of Scotland, a member of the Highland Society, a council member of the Wernerian Natural History Society, a founding director of the Astronomical Institution of Edinburgh, and a fellow of the Royal Society of Edinburgh, the Geological Society and the Society of Antiquaries of Scotland. He had a consuming interest in coastal erosion, and made a particular study of marine worms boring into submerged timbers, leading to the widespread adoption of the more resistant greenheart. While Robert, and indeed, all the Stevensons lectured widely, published many scientific papers and articles, the most notable of these was perhaps Robert's own description of building the Bell Rock.

Eleven miles out from Arbroath lies the most feared hazard to shipping on the east coast of Scotland – Inchcape Rock, nearly 300 m (1000 ft) long. At high tide it lies under up to 5 m (16 ft) of water. During the lowest of spring tides its highest point lies only a metre above the sea. Fully exposed it extends some 150 m by 80 m (160 by 87 ft) but at low neaps scarcely any part of it is visible. Lying in the path of ships plying up and down the coast from the ports of Aberdeen, the Tay and the Forth or across the North Sea to the Baltic, it presented a considerable hazard. Someone calculated it wrecked up to half a dozen ships every winter. It was here that Robert Stevenson's considerable reputation was first established.

Legend has it that in the 14th century the Abbot of Arbroath fixed a bell to Inchcape, which rung to the motion of the wind and waves to warn mariners of impending danger. The tradition then states that a pirate, who had long profited from the wrecks, tore down the bell, then famously perished upon it. A popular 19th century poem by Robert Southey called him Ralph the Rover:

> *Sir Ralph the Rover tore his hair,*
> *And cursed himself in his despair.*
> *The waves rush in on every side;*
> *The ship sinks fast beneath the tide …*

Bell Rock, off Arbroath

> *For even in his dying fear*
> *That dreadful sound assails his ear,*
> *As if below, with the Inchcape Bell,*
> *The devil rang his funeral knell.*

Thus this notorious place became known as the Bell Rock.

In the summer of 1794, whilst on a voyage to the Northern Lighthouses, Stevenson had changed course to have a closer look at the Bell Rock. Observing how the sea crashed upon it he wrote: '... from this period, the difficulties which must attend the erection of a habitation on this rock, appeared in a stronger point of view than they had hitherto done. Nevertheless, [I] resolved to embrace every opportunity of forwarding this great object.'

The Commissioners of Northern Lighthouses were also well aware of the hazard but were at a loss to know how to deal with it. They had considered the project in the 1790s but opted for Pentland Skerries instead. During storms in December 1799 over 70 ships foundered along the Scottish coast, two of them on the Bell Rock, so the Commissioners were forced to grasp the proverbial nettle. This time they charged Robert Stevenson to survey the rock.

After considerable difficulties, '... the crew being unwilling to risk their boat into any of the creeks in the rock', he finally managed to effect a landing, on 5 October 1800, and remained for two or three hours. He confessed: 'I will venture to say there is not a more dangerous situation upon the whole coasts of the Kingdom, or none that calls more loudly to be done than the Cape or Bell Rock.' Still reflecting upon the dilemma he later confessed to himself: 'I am sure no one was fonder of his own work than I was until I saw the Bell Rock.'

In January 1804 HMS *York* was lost with all hands at the Bell Rock. Several floating buoys were installed over the reef but none had survived the winter storms. Passing the Bell Rock from time to time had not lessened Stevenson's resolve. In 1803 he pondered: 'The more I see of the Rock the less I think of the difficulty I at first conceived of erecting a building of stone upon it.'

He remained convinced that only a stout stone structure similar to that devised by Smeaton for Eddystone was required, the workforce being accommodated on temporary barracks erected alongside. However since none of the rock showed above the high tide there would be additional technical challenges never faced by Smeaton. Two great engineers of the day, Thomas Telford (who failed to land in 1800) and then John Rennie (who visited with Stevenson in 1805) had both been asked to tender. Rennie agreed with Stevenson that a Smeaton model was the only option and dismissed Telford's estimate as impossibly low. He was however somewhat apprehensive about tackling a lighthouse, let alone on Bell Rock, but estimated the cost as only slightly less than one reckoned by Stevenson.

The NLB had considered 35-year-old Stevenson still to be lacking in experience but in 1806 agreed that he should work alongside Rennie as assistant engineer. In the end Rennie only achieved three or four brief visits to the rock in four years and Stevenson was not only left to his own devices, but was able to ignore most of Rennie's recommendations! Now, suitably impressed by Stevenson's abilities, the NLB were ready to accept him as their chief engineer in 1808 (a post he retained until 1843) and history now records Robert Stevenson alone as the builder of the Bell Rock lighthouse. Roland Paxton (2011) presents an interesting take on this awkward situation.

Work began at the Bell Rock on 16 August 1807. First, an 82 t fishing vessel, converted and renamed the *Pharos*, was put in place to act as a floating beacon or lightship and to accommodate the workforce until a wooden barrack was erected on the rock. Night and day during low tide, and only when the weather permitted, a foundation 14 m (46 ft) in diameter was gouged out of the bedrock, along with the foundations for the barracks. Although deeply religious, Stevenson sanctioned Sunday working since those willing to accept the small bonus were involved in a Christian mission to save lives. One-ton blocks of granite were cut from Rubislaw quarry near Aberdeen, to be shaped at Arbroath, and dragged to the dock (by a horse known as Bassey) for shipping out on two tenders named *Smeaton* and *Sir Joseph Banks*. The blocks were shaped to interlock like a three-dimensional jigsaw, just as Smeaton had done, each storey keyed in with wooden trenails. Stevenson established a forge in the barracks to sharpen the tools and even laid a small gauge rail track out on the rock to move the blocks into position. The tower would taper to only 4 m (13 ft) in diameter, its upper storeys being constructed of local sandstone.

Constructing the Bell Rock, model in the National Museum of Scotland

Up to 50 men might be working on the reef each season from late April or May through to October. Three hours was deemed a good working shift before the tide forced a retreat back to the *Smeaton*. The foundation pit then had to be pumped out again before work could begin on the next tide. The *Smeaton* and one of the three rowing boats broke their moorings, leaving 32 men (including Stevenson) stranded on the flooding rock without enough boats for all to escape. Fortunately, just in time, a pilot boat arrived from Arbroath, delivering mail. The *Pharos* broke her mooring just three days later but no men were on the rock. In that first season a total of 180 hours only were spent at work on the rock. In 1808, with the barrack completed, 265 hours resulted in 400 blocks being laid up to a height of nearly 2 m (6 ft), and by July the following year at twice that height the tide was no longer covering the structure. As a result more time could be spent at work, with the men safely barracked in the tower itself. In all 2,000 tons of stone were assembled, bound together using Smeaton's recipe for marine cement. The whole structure, 36 m (118 ft) tall, was completed in 1810. Then 24 Argand oil lamps were installed in the lightroom, each with a two-foot reflector of silver-plated copper. Finally, the keepers remained on station to illuminate the light, a distinctive red and white signature visible up to 35 miles (56 km) and lit on 1 February 1811. A final touch was the addition of two large 5 cwt fog bells as if to cock a final snook at old Ralph the Rover!

The whole venture had taken four years to build at a cost of £61,331. Admittedly, this was £20,000 over budget – but consider the significant dangers and harsh conditions under which the men had laboured.

In the days before radio or telephone, the keepers at Bell Rock – and indeed other pillar lighthouses to come – maintained communication with the shore station at Arbroath by a special code of signals. This assumed of course that the lighthouse was not obscured by fog or haze. Two long poles protruded from each side of the lantern gallery from which large black discs could be suspended, up to two on each side in any prearranged combination depending upon the message. Looking out one day in the early 1900s the keeper on shore spied a message, which he interpreted as 'Send Boat'. That evening the duty keepers, including John Maclean Campbell (of whom more later), were surprised to find the relief tug arriving without warning, but what had been misread as a desperate call for help turned out to have been merely two shags or cormorants that had been perching on the pole, inadvertently posing as the emergency signal!

Withstanding regular winter storms and occasional bird strikes (hence the net around the lamp room nowadays), the lantern and living accommodation has undergone several upgrades. The new tonite fog signal (finally replacing the bells) exploded in 1890 extinguishing the light for the first time ever, for over a whole week. In common with many lighthouses, the Bell Rock light was extinguished during the First World War but could be lit on request. On 27 October 1915 the radio message never got through leading to the foundering of HMS *Argyll*; all 655 crew were saved. An enemy shell blew up on the reef outside during the Second World War when the tower was also strafed by machine gun fire several times. In 1955 a helicopter, on a training mission from Leuchars and diverting to drop newspapers for the keepers, crashed below the light killing the crew. The tower

Bell Rock signal station in Arbroath

Bell Rock at sunset

experienced a disastrous fire in 1987 just before being automated in 1988. The lightkeepers are now gone, yet the Bell Rock lighthouse still stands today, 200 years on – the oldest working rock lighthouse anywhere in the world.

On his annual inspection tour of 1814, Sir Walter Scott accompanied Stevenson. When they visited the Bell Rock on 30 July the eminent author penned the following in the visitors' book:

Pharos loquitur
Far in the bosom of the deep,
O'er these wild shelves my watch I keep;
 A ruddy gem of changeful light,
Bound on the dusky brow of night,
The seaman bids my lustre hail,
And scorns to strike his timorous sail.

It was then too that Stevenson decided that his masterpiece be painted white. In 2006 Christopher Nicholson would conclude: 'If [Robert Stevenson] could be remembered for just one achievement then it would surely be this lighthouse. Its defiance of the waves is matched only by the dedication of its builder.' RW Munro was more practical in his assessment: 'Those who know it most intimately say that the best view of the Bell Rock is over the stern of a ship!'

Six

An Endless Calendar

*The usual signs which to the landsman's eye chronicle the passing
seasons are here unknown; but to us, the fish, shell fish, marine plants,
migratory birds, etc., constitute an endless calendar.*

John M. Campbell (assistant lightkeeper, Bell Rock), 1904

No account demonstrates the enquiring mind and literary abilities of lightkeepers better than John Maclean Campbell's *Notes on the Natural History of the Bell Rock*, a series of articles penned for the local press (*The Arbroath Guide* and *Chamber's Journal*) to be published as a small book by David Douglas, Edinburgh, in 1904. Although long out of print, and hard to find second hand, I discovered an online version, which would be printed and bound (as a softback) on demand. It proved quite a find.

Who would have thought that a granite tower 18 km offshore and surrounded only by a few hundred square feet of encrusted sandstone exposed only for a few hours each tide, and accessible for exploration only on calm days, could furnish – under Campbell's pen – such a wealth of careful natural history observation? I make no apology for delving into and quoting at length this little known, and hard to come by, literary treasure. Not only does it prove to be a fascinating account for the naturalist but it also provides a flavour of what it was like to be isolated on a wave-washed vertical prison for weeks on end, albeit with regular breaks ashore as soon as the elements allowed, for – in Campbell's case – nearly nine years. His narrative covers the period from April 1901 to April 1904.

Campbell was born in 1861, somewhere in Scotland although I have not yet discovered where. From his *Notes* we can infer he spent some time as a deep-sea sailor. One of his first postings with the NLB – presumably as a supernumerary lightkeeper – was at a lighthouse in Orkney. On 17 June 1895 he was posted to the Bell Rock as assistant lightkeeper where he spent 8 years 10 ½ months. The Bell Rock was not a popular posting and seems to have suffered a high turn-around of keepers. But there were some who actually relished the lonely life, and Campbell seemed to be one of those.

According to a brief obituary in the unlikely pages of a North of England natural history journal called *The Naturalist* (January 1917), written by Riley Fortune, Campbell was:

Left: John Maclean Campbell, 1861–1916

Right: The Bell Rock lighthouse

… a naturalist of the best type. A keen, careful and original observer, his positions gave him unique opportunities of studying nature at first hand. Whilst stationed on the Bell Rock Lighthouse, he published a charming little work upon *The Natural History of the Bell Rock*. A somewhat amusing incident occurred with regard to this. On one of the visits paid to the Bass [Rock] by the writer, he was comparing notes with Mr Campbell upon certain natural history books, when he advised him to read an interesting little book written by a fellow lighthouse keeper on the Bell Rock. 'Do you remember his name?' asked Campbell. 'No, I am sorry to say I do not.' 'Well, it was I!' This was naturally a great surprise, but a pleasant one, as it gave an opportunity of expressing sincere congratulation to the writer.

His Bell Rock narrative occasionally digresses into observations on daily life and work at the lighthouse, and indeed its history. In August 1902, for instance, with the aid of a telescope, he and his companions were even able to watch the Coronation celebrations in Arbroath. Understandably, Campbell was preoccupied with the weather but his principal subject was the natural history – somewhat restricted, one might think, in such a remote and sparse environment. But no, little escaped his enquiring mind.

He begins: 'The usual signs which to the landsman's eye chronicle the passing seasons are here unknown; but to us, the fish, shell fish, marine plants, migratory birds, etc., constitute an endless calendar.'

In January 1902 he explained:

The Bell Rock is a reef of red sandstone under 10–12 ft of water at high tide. When the sea ebbed however a rock platform no more than 400 by 250 ft in extent was exposed for a few hours. On calm days this afforded a welcome if slippery relief for the keepers, confined within the tower for weeks on end … A ramble round the rocks at low water just now discloses a scene of bareness quite in keeping with the season of the year. The upper surface of the higher lying rocks is as bare as a street pavement, and only an occasional patch of acorn barnacles remains of the encrustation with which they were invested during the summer … Vegetation now exists only at low-water mark; above that, broken tangle roots, or, to be more correct, the claspers are seen still adhering to the rocks, the tangles themselves having been shorn clean from their moorings …

In stark contrast to the barren and exposed rock of, say, Dubh Artach, Bell Rock seems to have offered a fertile field of enquiry for a keeper of Campbell's capability. Rather than follow his *Notes* chronologically, from his first jottings in April 1901 to the last three years' later, I shall reconstruct his account into a representative calendar year.

The Bell Rock landing stage at half-tide

JANUARY

1902

To witness the continual thumping and pounding to which the Rock is subjected during the winter, one is surprised to find that life in any form should continue to exist under such conditions … A close search reveals exceedingly minute forms of life. Here in this stony basin … is a small crab, so small indeed that a split pea might easily conceal him. He is not a younger either, but fully adult, in proof of which we have frequently found them, in the proper season, with their spawn attached. Deep in his little pit he seems quite immune from the furious seas that tumble overhead as the tide makes … The white whelk, so much in evidence here, have all gone into winter quarters, and underneath projecting ledges and in sheltered nooks they may be seen in myriads, their position being so judiciously chosen as to be completely protected from the heavy north-east seas.

Over the recently emptied contents from the cook's slop-pail a flock of gulls are circling and screaming, actually hustling each other in their attempts to capture anything edible. A solitary 'black-back' is seen among the noisy crowd, and, as he swoops at some tempting morsel, his black, beady eye watches our every movement with suspicion. What a handsome bird he is as he swings past within a few feet of us, the back and wings presenting a dead black appearance in startling contrast with the immaculate whiteness of the fan-shaped tail and the remainder of the body. Despite his handsome appearance, he is a veritable vulture, and nothing comes amiss to him in the way of food, be it fish, flesh or fowl.

Sea pheasant is the name by which the long-tailed duck is known in some localities, and as we watch a flock of them crossing the reef in full flight, the synonym is at once apparent. In style of flight and shape, to the long tail feathers, they are similar to the pheasant, but only half the size, with beautiful plumage of black and white. Here they are known as 'candlewicks', their call notes needing but little stretch of the imagination to be rendered 'Here's a candlewick', repeated several times in shrill falsetto, which on a quiet day becomes somewhat annoying as it clamorously floats through our bedroom window.

Some queer visitors we have here at times in the way of birds. Once we captured a large owl dosing sleepily in one of our windows. During the week of his captivity he would not deign to partake of any food we offered him. Coming off watch one night I took one of a flock of larks which were making suicidal attempts to pierce the plate glass of the lantern. Placing it in the room where the owl was roosting, it fluttered to the window, when, like a flash of lightning and equally as noiseless, from the other side of the room the owl came crash against the glass, a few feathers later on testifying his appreciation of this form of dietary.

1903

Bright, sunny weather characterised the opening day of the year, the sea assuming a suspicious placidity quite summer-like in appearance but for the keen nip in the air perceptible out of doors. This state of affairs, however, proved but ephemeral, and for the remainder of the month we have experienced most boisterous weather.

Left: Herring gull
Right: Gannet in flight

1904

The weather continues dull and dark, but comparatively quiet – a matter of much importance to us at relief times. We have no aversion to a rousing gale between reliefs; then one can afford to appreciate the grandeur of the warring elements ... A solitary lapwing was our only feathered visitor for the month.

FEBRUARY

1902

Piercing cold weather here of late, with a good deal of frost and occasional snow showers. No matter how heavy the snowfall may be here we only see it falling, as it does not lie long round *our* doors, and only when our gaze is directed Arbroathwards – which, you may be sure, is not seldom – are we reminded of its occurrence. The close of last month saw our barometer taxed to its utmost intelligence, and though a tenth higher would have seen its limit, nothing of a phenomenal nature was noted.

A solitary [gannet] was seen in the first week of December, and since then the number sighted has gradually increased, till in the middle of the present month, as many as eight in one string were counted winging their way southward ... February (1903) has been a repetition of [last month] – cold and blowy, with excessive rainfalls. The first week saw hundreds of [gannets] back to their breeding haunts on the Bass Rock. In 1904 [Gannets] arrived at their breeding haunts on the Bass Rock from their southern sojourn on the 9th of [January].

In a shallow depression on the higher rock surface our attention has been attracted to a solitary plant, a specimen, I understand, of *Himanthalia lorea* ...

1904

Cormorants have been more in evidence here this month than usual. At present a flock of thirteen is to be seen diving in the deep water surrounding the reef … On several occasions during the month a seal was observed sporting amongst the breakers. The other evening he was within a few yards of the tower, busy devouring a huge cod … Our feathered visitors for the month were represented by a couple of skylarks, three song thrushes, a pair of carrion crows, and a solitary starling. Eiders and long-tailed [ducks] still continue in attendance, and gannets are now plentiful. The month [February] has been wet, cold, and stormy, exceptionally heavy seas prevailing in the earlier half. The closing day of the month was beautifully clear and sunny, but cold and frosty, our heliograph [contact with the signal station in Arbroath] intimating on that date a similar state of weather on shore.

Left: Eider pair display
Right: Long-tailed ducks

MARCH

1902

Signs of uneasiness and unrest are now apparent amongst our winter boarders, the eiders and long-tailed ducks. Taking wing on the slightest provocation, they wheel aimlessly round the Rock, and instead of their usual steady persistence in diving for a living, they seem quite discontented with their lot, and plainly making up their minds to desert us for the summer. Advances by the males are as yet met with scornful rebuffs by their less showy plumaged partners, but soon a mutual understanding will be arrived at, and before the month closes they will have gone house-hunting, eiders possibly to the Isle of May, while the long-tails, being migratory, seek their homes in the frozen North. It seems a strange anomaly that the less robust looking long-tail should choose such rigorous latitudes for the rearing of its brood, while the sturdy 'dunter' [Eider] swathed in his Arctic coat, should elect to stay at home. On the other hand, we have been visited on hazy nights by numbers of larks and thrushes returning to our shores, after wintering in 'Norroway over the foam'. These members of the spring migratory movement often come to grief on our lantern, and when one considers the number of lighthouses round our coasts, it will be understood that the death-toll from this cause alone must be extremely high. Designed to save life, we unwittingly lure our feathered friends to their destruction.

A couple of seals have been sporting round our door of late, and they also exhibit signs of exuberance in keeping with the season.

The rocks become at this season of the year invested with a slippery coating of algae, which renders it extremely difficult to maintain one's footing, and also necessitates repeated applications of hot lime to our gratings in order to render them passable. Myriads of minute whelks, no larger than turnip seed, strew the rocks and crunch under foot as we walk, while great patches of mussel spawn delight the heart of the more venturesome of the white whelks … Fishing in the Rock pools has been tried for the first time this season, and resulted in the capture of a solitary 'cobbler'. It may be a month hence before we meet with any success.

1903

Borne on the off-shore wind comes the odour of heather – not the fragrant perfume one usually associates with this sweet-smelling plant, but the smoky incense consequent on moor-burning …

… The rocks are this year more plentifully strewn with mussel-spawn and acorn barnacles than usual; and already the whelks have sallied from their winter's sleep, bent on their destruction. Hundreds of hermit crabs have also made their appearance, notably first in the deeper pools, but gradually taking up their quarters in the shallows.' Not surprisingly, when Stevenson's work force were constructing the Bell Rock lighthouse between 1807 and 1811, they became intimate with every rock, ledge, crevice and pool. All were carefully mapped and named – some 70 in all – which Robert included on a detailed engraved map at the end of his *Account*.

As he scrambled around the reef Campbell became particularly fascinated by lumpsuckers or, as he called them, paidlefish or poddlies. In January 1903 for instance, he observed how:

… a round of the different fishing pools at low water resulted in the capture of a most unhealthy-looking specimen of a poddley in the 'Hospital' (Neill's) Pool … It is most remarkable why these sickly fish should favour this pool alone … Possibly its greater depth offers a safer refuge for these convalescents than the other pools. The surface of this pool measures about three fathoms across, and a fathom and a half in depth, when the tide leaves the Rock. The bottom is generally covered with boulder-stones, which are whirled about with much force when the sea is in a state of agitation. [According to the Bell Rock website Mr Patrick Neill was a friend of Robert Stevenson and in the summer of 1808 landed at the Rock to examine the seaweeds and marine creatures and the Pool was named after him].

Towards the end of [March 1903], a few small spats of paidlefish spawn were seen deposited in convenient crevices of the rocks. This is unusually early for 'nesting' operations, and engenders hopes of an early fishing, as the ova are generally the first inducement for the wandering cod to come within reach of our rods ... Numbers of peculiar looking slugs are met with at present, somewhat resembling a section of a small orange with the skin attached ... Another small slug noticed this month – no larger than one's finger nail and recalling the general appearance of the 'fretful porcupine', with 'quills' arranged along its back, and displaying beautiful shades of brilliant blue and crimson ...

Saturday 14th – A beautiful warm sunny day, the sea like glass, dappled here and there with great greasy-like patches peculiar to still weather. Flocks of eiders, long-tails and gulls appear to be having a day off, and float listlessly hither and thither, seeming only intent on making themselves aggressively audible in the stillness, the long-tails piping a shrill treble to the sonorous bass of the eiders, while the gulls contribute a fairly good imitation of a laughing chorus ... But few spring migrants have come our way this month, principally a few blackbirds, thrushes and starlings.

With the exception of a few days, the weather [March 1904] has been extremely favourable; indeed, for the greater part, summer-like – a pleasant change from what we have experienced of late. The peculiar white rubber-like folds of ribbon which have been adhering to the Rock surface for the past two months, and which we erroneously supposed to be the ova of some fish, turn out to be the spawn of the slugs I have already described ... These shell-less molluscs have been much in evidence this season; and representatives of three distinct families are to be met with, namely the Onchidoridae, Tritoniidae and Eolididae. Cannibals, they attack their own species without compunction, and devour each other's spawn. Darwin computed that some 'ribbons' contained as many as six hundred thousand eggs. The acorn barnacles which have escaped the voracity of the white whelks have in some places attained a height of two inches ... Hermit crabs are at present abundant ... starfishes – principally the five-rayed variety – are now numerous, and garnish each shallow pool. Sea-urchins, though never plentiful here, are occasionally met with, some having been found recently no larger than a pea.

On the 20th the advent of the paidle-fish was announced by a small patch of ova underneath a projecting ledge of rock, and, on the same day, by a reconnoitring 'cock'. The young of last summer are met with adhering to stones in the shallow pools; and, contrary to our expectations, though only two inches long, were found to contain spawn.

The spring migratory movement has sent but few birds our way this year. A few thrushes, blackbirds, larks, starling, and a couple of pied wagtails composed our list.

Left: Song thrush

Right: Pied wagtail at glass pane

By the middle of the month, the long-tailed ducks had gone north to nest, but four pairs of eider now remain.

APRIL

Early in April 1901

… the gulls, which have been levying blackmail from the ducks all winter, have almost all disappeared, and we miss their raucous voices at our door, contending for after dinner scraps.

Occasionally a profuse spat of mussels might appear but Campbell considered the depredations of the whelks to prevent them from successfully colonising. He noted how some workmen engaged in the construction of the lighthouse 100 years earlier had transported mussels out to the rock in the hope of adding fresh shell fish to their otherwise monotonous diet, but they too failed to establish.

It is testament to the curiosity of the bored keeper that Campbell experimented with the hermit crabs he sometimes found in whelk shells:

An amusing sight may be witnessed by placing several of them deprived of their shells, in an ordinary soup plate, with a little seawater and some empty shells – fewer shells than crabs. The fighting and struggling to secure houses is ludicrous in the extreme.

The extremely low tides prevalent at the opening of [April1902] enabled us to extend our hunting grounds somewhat further than usual … Quite a forest of luxuriant tangles now cover the lower lying portion of the reef. Their dripping blades appear on the surface, scintillating in the brilliant sunshine like so many diamonds, till the receding tide permits the warm sun to rob them of their freshness, their beauty vanishing in a perceptible vapour, leaving them flaccid and inert till the returning tide restores their pristine beauty. The badderlock or benware is here also in great profusion, and usually selects a position the reverse of peaceful, being generally found where the wash of the seas is most constant. Of rapid growth they attain a great length, some measuring fully sixteen feet; one we had under observation was seen to increase a foot in length in six weeks.

Owing to hazy weather we had a number of compulsory visitors to dinner yesterday. Seated outside our kitchen window was a party of fog-bound travellers, consisting of a pigeon, a starling, a wagtail, a robin, and a couple of wheatears. The starling was sitting bunched up by himself, preserving a stolid indifference at his enforced detention, and appeared to treat the animated expansion and flirting of the wheatears' tails as undue levity, unbecoming their sorrowful predicament. The beautiful black-throated wagtail is all alertness, and the slightest movement on our part sends him circling round the Rock till, unable to sight the land, he is fain to regain his resting place.

The pigeon has been here a week now, and evidently has no intention of leaving. Should the window be left open he makes bold enough to enter, although but the other day he gave us a somewhat dramatic illustration of the proverbial hen on the hot 'griddle' by rehearsing a fandango on top of our cooking range, a position from which he had to be forcibly removed. Today, the 21st, he has been joined by a companion of his own species, a red-chequered homer; but instead of the mutual demonstrations of pleasure one would expect to witness at their meeting in such isolation, they remained quite indifferent to each other's presence, the newcomer possibly from motives of disdain, as he appears to belong to the aristocracy, seeing he sports an aluminium bracelet, on which are the letters 'UB' and the year '1901', besides a number composed of three figures, which, unfortunately I took no note of. A strong southerly breeze on the 22nd deprived us of their company. Losing the shelter of the tower, they were unable to make headway against the wind, and, fortunately for themselves, were driven landwards.

A few months later, Campbell was careful to record the inscription on the rings of two other homing pigeons, finding that one had come from Leeds, and the other from Beeston [probably the one near Leeds]. 'After being watered and fed, both were released at 11 am on 13th July, each steering a sou'westerly course away from the rock,' – presumably on their way home.

1902

On the 20th April 1902 a small patch of paidle-fish [lumpsucker] spawn was seen cemented in a sheltered nook of the rocks. This is unusually early for nesting operations, as it is generally May before they are much in evidence here … a hopeful sign, and is the first attraction for the wandering cod, by whom it is greedily devoured, providing they can steal a march on the red-coated sentry – a difficult matter, one would think, considering how assiduous he is in the protection of his charge.

Paidle-fish are now fairly numerous, their nests, with attendant cocks, being met with on every hand ... Stretched along the rock, my face close to the surface of the pool, I had an excellent view of the nest and its guardian, two feet below.

Campbell had watched this male lumpsucker for a while. Defending its eggs, the fish had grabbed a predatory white whelk in its teeth, carried it a couple of feet to the surface of the water and spat the whelk out, almost in Campbell's face! He writes, 'Since then I have repeatedly seen them remove predatory starfishes and whelks in a somewhat similar manner.'

Left: Dog whelks spawning
Right: Lumpsuckers (Harald Misund)

1903

The wheat-like ova of the white whelks is also large in evidence this month, though somewhat later than last year ... A stranded cuttlefish was an object of much interest one evening. What a queer-looking object it appeared, with its eight long tentacles squirming in all directions, its body a slobbery mass of animated mucilage. Although only a foot in diameter it required some force to detach it from the rock ... Irritated, it appears to have the properties of the chameleon, flushing through all the gradations of colour in quick succession, and latterly discharging a jet of fluid of inky blackness ... Frequent irritation, however, exhausted its stock of ink, and latterly only clear water was expelled. This expulsion, when affected on the Rock, was accompanied by an audible murmur. The narrow slits of eyes closely resemble those of a dog-fish, and the head, with the anterior tentacle elevated in the air, grotesquely reminds one of an elephant in the act of trumpeting.

Warm, sunny weather in the earlier part of [April] … A few gulls, herring, and kittiwakes hover about, and guillemots and gannets are now common. The gannets, I am informed by the keepers on the Bass Rock, commenced laying there on the 11th … In 1904 the last four pairs of eiders took their departure on the 17th April, and only a few gulls now remained.

MAY

1901

Flowery May! Well, not exactly. To us this month generally spells fish … though planted here, right in the centre of the fishing grounds, our table for the greater part of the year is 'fishless'; for unlike Mahomet and the mountain, the fish must come to us, even to our very door – for from our doorway [fifteen feet above the sea] the most of our fish are caught, save an occasional one taken with rod and line from the deep pools left on the Rock by the receding tide … Strangers often ask why we do not keep a boat here; it might almost as reasonably be asked why we don't keep a cow. Simply because we have not the necessary accommodation – that is, unless [a boat] could be devised with the properties of a limpet, and be none the worse for several hours' immersion every tide.

Early in May lump-suckers visit the Rock to deposit their eggs, cemented in a compact, gelatinous mass to the rock. The hen is about eighteen inches long, of a somewhat repulsive appearance. The cock is about half this size, and more attractive, being brilliantly coloured, combining various shades of blue, purple, and rich orange. A broad sucking disc between the pectoral fins enables the fish to moor itself to the rock and maintain an upright position … This operation performed, the hen evidently considers her share of the contract as finished, as she immediately clears out to deep water, leaving the cock to mount guard over the nest … I have frequently taken them away from the nest and placed them in a different part of the pool, but they invariably returned to their post.

Careful observation – even elementary experimentation – of marine life was providing Campbell, if not his companions as well, with hours of amusement. With sorties abroad on the rock being limited by tide and weather, it is not surprising that they desired a closer look so in May 1901 they embarked on an ambitious little project.

We have just completed a small aquarium, by means of which we hope to become better acquainted with the more minute organisms with which the Rock at this season of the year teems. Apart from the study of man himself, what can be more interesting than to be an actual eye-witness of the gradual evolution of the different forms in the great life-scheme of the Creator, from the simple nucleated speck of protoplasm (amoeba), which multiplies by simple division, to the more complex structure of the members of the vertebrate kingdom?

In June 1901, he added:

… the other day while collecting specimens for our aquarium [we found] a spider-crab which, when stranded high and dry, appeared but an unsightly mass of tangled seaweed, [but] when placed in the aquarium assumed all the beauties of a verdant grove. From every available point on the upper surface of his mossy-covered shell beautiful variegated plants streamed and waved their delicately-feathered fronds. Conspicuous amongst this luxurious growth were specimens of the corallines or sea-firs, which a casual observer might easily suppose to be miniature fir trees, but which in reality are plants only in semblance. Each of these delicate looking plants is actually an animal; in fact, a colony of animals. … In the second instance our attention was centred in a small jelly-fish swimming about in a quiet pool, its translucent body being scarcely distinguishable but for the beautiful flashes of iridescent light it was continually giving off in the bright sunshine …

A chip of rock covered with acorn barnacles becomes an interesting object when placed in the aquarium. Each conical shell is packed as close as possible to its neighbour, apex upwards. The apex is open, and fitted with a lid composed of four shells. Under water these lid-shells are seen to separate, and a bunch of 'fingers' set on a stalk are thrust out, make a clutch, and are withdrawn. The 'fingers' have extremely fine hair-like processes fixed at right angles to them, the whole forming a sort of net through which the water is filtered and the minute food-forms retained. It is interesting to know that although now fixed immovably to the rock these animals began life as free swimmers, and, strange to say, closely resembled the young crab.

Another object we had under observation at the same time was one of the sea anemones, named the dahlia wartlet [*Urtica felina*]. A fleshy-looking disc studded

Left: Rock pool and anemone
Right: Arctic tern

with pieces of broken shell and sand, it appeared anything but attractive; but seen in the aquarium the connection with its floral namesake was at once apparent. Unfolding itself from an orifice in the centre … rows of beautiful coloured tentacles were disclosed. These tentacles have the property of adhering to any object they come in contact with, and contain within themselves some wonderful mechanism. Placing a fly on the extremity of one of the tentacles, it was immediately held fast. The whole of the tentacles then curled inwards, carrying the fly with them, thus clearly showing their function.

December in May fitly describes the prevailing state of the weather [in 1902]. Chilling winds, accompanied by snow, hail or sleet showers, engender doubts as to the veracity of the calendar, but the arrival of a number of terns on the 18th dispels all doubts upon the matter. Sojourning in Africa since their departure in September, they invariably make their appearance here in May. At present there are about thirty of these energetic little birds busy diving among the breakers, picking up small fry, among which is seen inch long sand eels. A flock of kittiwake gulls also hunt alongside of them, while several gannets are to be seen further off, plunging in pursuit of larger game.

Clustered in sheltering nooks of the rock are numerous patches of ova, deposited by the white whelk. Closely resembling ears of wheat in size and shape, each is attached to the rock by a short footstalk, terminating in a flattened disc. On being pressed, a milky fluid, somewhat granular, is exuded from the free end. The whelks themselves are at present feasting on limpets, whose shells have been fractured by the debris consequent on the alterations in progress [a work party were replacing the lantern in the lighthouse], though at other times they do not appear to attack the limpets, their thick shells possibly making the game not worth the candle.

During the first few hours of [May 1903] our lantern was the centre of a twittering throng of feathered migrants. Wheatears, rock pipits, starling, wrens, and robins fluttered erratically through the rays or clamoured in the innocence against the glass, apparently desiring a closer acquaintance with the source of light. Puffs of feathers floated away on the easterly breeze as some unfortunate, less discreet than his fellows, crashed against the invisible barrier. The coming dawn, however, reveals to the survivors the absurdity of their position, and ere the light is extinguished they have resumed their journey shorewards.

Frequent fogs occurred in the earlier part of the month, and during the prevalence of a long spell a long-eared owl was captured on the balcony and held prisoner for a week, during which time various samples of our commissariat were offered for his acceptance without avail. A luckless sparrow, the only one by the way I have seen here, was then captured and placed at his disposal. This proved more in his line of business, for on the morning after the rump and tail feathers alone were left. Next day the indigestible portions, feathers, etc., were cast up in the form of a compact ball. Later a thrush was similarly offered, but after a couple of days in each other's

Left: Wheatear

Right: Lion's mane jellyfish

company remained untouched. It was amusing to see the spirited attitude assumed by the thrush when in the presence of his natural foe. Screaming aggressively at the slightest movement of the owl, he would lunge furiously in his direction, his bill all the while snapping audibly. The fog having cleared somewhat, both were then set at liberty.

Another very rare visitor seen here this month was a Sheldrake, which passed close overhead flying south. This is the first I have seen here … Just as the rocks were being overflowed the other day, we had a visit of another bird which is but rarely seen here, namely the oystercatcher … Wading an inch or so deep, where the limpets were probably opening to the influence of the incoming tide, he appeared to make a judicious selection …

JUNE

1901

A heavy, pounding, nor'-east swell in the early part of this month has almost denuded the higher portions of the Rock of the young vegetation, the tangles being as cleanly cut as if by a reaper. The [lumpsucker] nests also suffered, as not a vestige of them was left. Several cocks have been seen wandering aimlessly about, their occupation evidently gone.

1902

Only towards the end of this month did we experience anything like summer weather. Belieing the wintry weather we had been experiencing, the fragrant odour of the hawthorn blossom borne on the off-shore wind imparts a pleasurable sensation… Scarcely a bird is to be seen in our vicinity at present, nesting operations calling them elsewhere. A few foraging gannets are seen daily passing and repassing, catering for their sitting mates on the Bass Rock.

JULY

1901

The few boats which have been prosecuting the lobster fishing here for the last three months have now abandoned it … A return of four lobsters for the hauling of fifty 'sunks' is but a precarious living … It is only within this year or two that the Rock has again attracted the attention of the lobster fisher, after a lapse of many years. Prior to that time, we could always rely on an occasional lobster being found in the holes on the Rock at low water; while crabs, which could be had in abundance are now extremely scarce, and the lobster, as far as we are concerned, might well be as extinct as the Dodo.. . We had a rather surprising catch in a lobster-creel one time here … a huge conger eel … A further surprise, however, awaited us; for, on being cut open, a full-grown lobster was found in his stomach.

The terns increased to over a hundred 'from daylight to dark their creaking voices are dinning in our ears.' Feeding chiefly on herring fry near the surface, the parents were still feeding some of the young birds, which had not yet attained their forked tail.

1902

Myriads of medusa or jellyfishes are constantly streaming past our door, apparently without any powers of volition of their own, but helplessly at the mercy of the tides. Of various sizes, shapes, and colours, they impart quite a gay appearance to the seascape, somewhat resembling a grassy sward carpeted with beautiful flowers … Great tremulous discs, twelve inches in diameter, trail their streaming tentacles several feet in diameter …

1903

The middle of July brought our first young tern, and towards the close of the month several were in attendance. A large school of bottle-nose whales crossed the reef on the last Sunday of July, their puffing and blowing being quite audible as they headed north, probably after herring.

AUGUST

1901

The extreme heat prevalent in the earlier part of this month was characterised by a number of unusual visitors but ill adapted for any lengthened sojourn here. Butterflies, bees, wasps, and moths innumerable were to be seen flitting about the rocks in the daytime and clinging to the lantern at night, numbers of them being drowned every time the rocks were submerged. Several daddy-longlegs and a single specimen of the beautiful painted butterfly were seen among the ill-fated host. Why should they have journeyed to such inhospitable quarters is not yet apparent, but possibly the steady westerly wind then blowing was responsible for their presence here, and, finding themselves unable to stem the current, like many unfortunates, they followed the line of least resistance …

1902

… several white butterflies and moths innumerable were seen passing here this month. It seems these insects have their migratory periods as well as birds, and at stations favourable for their observation they appeared, to quote from a writer in a recent number of *Chambers's* 'as a dense snowstorm driven by a light breeze, and this not for one day only, but for many in succession. Whereas birds come and go with clockwork regularity, the immigration of butterflies is uncertain, and of all those which survive the perils of the deep no single one returns.'

Painted lady butterfly

The sea in our vicinity is just now actually alive with shoals of immature fish, chiefly sand-eels, herring-fry, and what appears to us to be finger-long whitings. Incessant war is continually being waged upon them by the terns from above and the poddlies from below. Watch this particular shoal of 'fry' as it swims with the current past our door; notice the orderliness which prevails amongst them, as of a disciplined army. But from below the poddlies have sighted them, and swift as light they are amongst them with a deadly rush. Completely disorganised, the 'fry' scatter in every direction … But suddenly a flash of bronze in the bright sunshine betokens the leap of the lordly lythe, as he seizes his victim from amongst the attacking-force and as quickly returns to his lurking-place among the luxuriant tangles … The full-grown lythe may be truly termed the poor man's salmon, not from a food point of view – though in itself not to be despised – but as a source of sport.

In July 1902 Campbell noted how: 'the coating of acorn barnacles with which the higher surfaces of the Rock and also the base of the tower are whitened in summer is fast disappearing before the ravages of that ruthless destroyer the white whelk.'

1903

Campbell recorded an interesting specimen:

During the unusually low tides occurring in [July 1903] an opportunity was afforded of examining a most peculiar form of animal life of which I have nowhere seen any account. Attached to the rocks amongst groups of coralline this curious object has all the semblance of a bird's claw' it turned out to be a whale-louse.

This is a crustacean of the family Cyamidae, one of 30 or so species up to an inch long, that are parasitic on whales and dolphins, attaching to folds of skin or wounds to feed off algae or flaking skin. They are often specific to only one species of cetacean and it is extremely rare to find one unattached.

SEPTEMBER

1901

The rock has taken on quite a wintry appearance. The vegetation on the more exposed portions has entirely disappeared under the influence of the heavy seas experienced during the greater part of [September] ...
... the terns have deserted us, and, to complete the prospect, on the morning of the 19th we had the first visit of our winter boarders, the eider duck ... The heavy easterly surf has deprived us for the present of our fishing, forcing the fish off the rock to deeper water. There are evidently plenty about, as the gannets are to be seen busy diving in the vicinity.

1902

A good deal of heavy winds has been experienced on the Rock, and the stability of our new lantern subjected to a fair strain, though probably nothing to what it will have to encounter during the course of the winter ... We have had several takes of fish of late, though there seems to be a scarcity of 'fry' compared with last year, the absence of which probably accounts for the terns failing to call upon us with their young for a few weeks' feasting prior to commencing their migratory journey southwards. Gannets may be seen at present striking at fish within a few feet of our doorway, while a flock of young gulls hover expectantly, with feeble peeping cries anticipating the feast in store for them when the dinner scraps make their appearance.
Further off, [swim] a few eider ducks – who only arrived on the 25th, somewhat later than last year ... The eiders present are as yet all adult males, the females presumably still occupied with family cares ... On the 6th we had our first intimation of the autumnal migratory flight in the arrival of a flock of wheatears, accompanied

by a solitary wren. On the 27th several greenfinches, larks, and starling were making insane efforts to follow the line of most resistance, resulting in our new lantern receiving its first baptism of blood, as the glass next morning testified. Several porpoises are to be seen puffing and blowing a mile off, and on the 28th a school of 'finner' whales were seen heading north.

1903

A solitary eider arrived on the 20th September but soon numbered a hundred or so. 'The long-tails are still awanting to complete our list, but their advent may be looked for early next month … On 4th October an extremely rare visitor – here at least – made its appearance, namely the ring-ouzel, the first I have seen.

OCTOBER

1901

The flock of eider ducks which keep us company through the winter increases daily, and now numbers over thirty. Swimming and diving amongst the breakers from daylight till dark, it is astonishing how they escape being smashed on the bare rocks … They are sometimes quite close to the tower, and then we have an interesting view of their proceedings. The diving of one is generally the signal for the remainder to follow, and the whole flock may be clearly seen, a couple of fathoms down, scurrying over the rocks in eager quest of the different dainties on their menu, consisting chiefly of small crabs.

In October 1901 Campbell recorded how 'on a hazy moonless night, with a sou'easterly breeze and drizzling rain' brought the autumnal migratory flight:

Making straight for the light, they dash themselves against the heavy plate-glass of the lantern; many of them are thus killed and swept by the wind into the sea. Others, again, arrive with more caution, and though taken in the hand and thrown clear of the tower invariably return, and remain fluttering against the glass till daylight reveals to them the futility of their exertions in that direction. The most numerous of these visitors are the redwings and fieldfares, but blackbirds, larks, starlings, wheatears, finches, tits, etc., may be met with in the course of the season …

1902

We have had occasional visits of feathered migrants during this month, but it is a matter of remark that each year sees a decrease in the number of arrivals here. Probably the increased number of lights on our coast accounts for this diminution, some proving more attractive than ours. A few years ago it was quite on the cards at this season of the year – thanks to the migratory instinct – to have an additional course at dinner, to which fieldfares, blackbirds, and redwings were the voluntary contributors, and even at times the gamey woodcock 'graced the groaning board' … Birds generally arrive here in a fagged condition, and are easily captured. As an instance, a kestrel landed on

Left: Redwing
Right: Eider ducks in flight

our balcony railing during fog, and, despite the explosions of our fog-signal twenty feet overhead, tucked his head under his wing and fell sound asleep! Another arrival of note was a common blue pigeon, which, after a few hours' stay, surprised us by depositing an egg in our doorway!

Podley-fishing has been fairly successful during the month, and several codlings have been taken from the pools at low water … A new light is being completed on the Bass Rock, and on the first of December, yet another factor in our dwindling list of visitors will be in operation – ostensibly a lighthouse – but to our feathered friends, alas! a veritable slaughter-house.

1903

On the morning of the 24th October over two dozen tiny gold-crested wrens were circling round our lantern, jostling and tumbling over each other in frantic efforts to keep in line with the white flash, the red flash evidently having no attraction for them. A skylark and robin were also of the company, as well as several redwings. The robin always seems to have a truer sense of his position than any other of our visitors. While the others clamour futilely against the glass he maintains an aloofness and self-possession truly remarkable. His eyes seem to be everywhere, and only with difficulty and the exercise of a little strategy is his capture affected. Of course, our captures are but temporary, and merely for the sake of a few minutes' examination … On 29th October a flock of thirteen fieldfares passed at 9 am, flying towards Arbroath. This is the first arrival here of these birds, and earlier than usual. On examining a lark which had been killed on the lantern the other night, a small land shell was found adhering to the feathers on the under part of the body.

NOVEMBER

1901

Boisterous weather prevailing for the greater part of [November], we have been closely confined to the house. Our connection with the amphibian being so extremely remote completely disqualifies us from enjoying our usual 'constitutional', the grating, even at low water, being occasionally swept by the heavy seas. Our winter boarders, the eider ducks, have been reinforced, on the morning of the 14th – somewhat later than usual – by the arrival of a flock of long-tailed ducks. These, with the eiders, will keep us company till April again calls their attention to domestic affairs.

1902

Exceedingly stormy weather, with a prevalence of sou'easterly winds and heavy seas, has been our portion here, restricting our movements out of doors … About the beginning of the month we had a few feathered visitors, chiefly blackbirds, fieldfares, and starling. On the morning of the 5th several struck heavily on the lantern, but were swept away by the strong sou'east wind then blowing. The gannets have now all disappeared, none having been seen since the 27th. The eiders continue in close attendance and have had their numbers augmented by the arrival of the long-tails on the 9th, a week earlier than last year, thus completing our list of regular boarders for the winter. At 6 pm on the 13th we were privileged with the unusual spectacle of a lunar rainbow. The bow – a faint white are against the dark background – was distinctly visible in the nor'west, though, of course, void of the vivid colouring inseparable from its solar namesake.

1903

A flat calm. A pleasant change, indeed, after our recent experience, and one which has fortunately continued for the greater part of the month. Fish, which had maintained a safe distance during the turmoil of last month, now ventured within catching distance, and several good takes were to be had. After the middle of the month heavy seas again drove them out of reach into deep water … Our flock of eider ducks, much larger than it has been for several years back, now numbers 120. Amongst the smothering breakers they seem to be in their glory, and are busily engaged in clearing off the immature mussels that have escaped the voracity of the white whelks. On the 2nd, the first two long-tails were seen, exactly a week earlier than last year, but their numbers are being but tardily reinforced, as they only totalled six at the end of the month. Though the main body of the solan geese or gannets left their breeding haunts on the Bass Rock on the 5th of the month for the fishing grounds – in the Mediterranean it is said – occasional stragglers are still seen in our vicinity.

On the night of the 8th we had a few migrants on the lantern, ten blackbirds – three only of which were males – and three fieldfares. Several of them appeared much fatigued, and after a few preliminary hops around the lantern, settled down on the lee side to have a nap.' Jostling with each other for position and those ending up on the windy side were soon swept away into the darkness. 'The haze, responsible for their appearance here, cleared after midnight; before 3 am they had all resumed their journey shoreward. On the 20th, a pair of grey crows passed, going east, and on the 22nd a heron was seen travelling in the same direction. Again, on the night of the 27th, three hen blackbirds and a starling had the lantern all to themselves …

The white whelks have now gone into winter quarters, and only a few are to be seen lingering among patches of immature mussels. The black edible whelk, or periwinkle, whose vegetarian habits demand a more inshore life, is here conspicuous by its absence. Occasionally, during the summer months, a very close search reveals a few solitary specimens. Two different varieties of slugs have been much in evidence among the rocks here of late. One of them (*Doris coccinea*), resembling in shape and colour a section of an orange, I have already described; the other somewhat resembles the common snail. Furnished with anterior horns and fleshy spines, ranged along the back, it curls itself up when out of water like a hedgehog. Earlier in the season they were mostly of a beautiful bluish colour, now they appear quite red.

Cooped indoors so long, one is glad to take advantage of the quiet weather and the absence of the tide to enjoy a spin along the gratings, even though at night and in darkness.

One night, battling against the wind, Campbell described all the lighthouses to be seen on the horizon – the Isle of May, Bass Rock, the North Carr and Abertay lightships, Buddonness and Tayport, Montrose Ness, Tod Head, Barns Ness and St Abb's; it was rare to see them all. He mused how:

The presence of these lights makes our coasts as safely navigable by night as by day, and the demand is still for more – a fact which drew from a facetious old 'salt' the remark that 'sailors nowadays want a hand rail along the coast.'

DECEMBER

1901 was notable for its stormy weather. But in 1902 Campbell could appreciate how:
The eiders and longtails, with an unswerving attention to business, pursue their calling amid the hurly-burly of broken, tumbling seas – evidently little concerned whether the weather be fair or foul … Gannets this month are conspicuous by their absence, and only a few parasitic gulls divide their attention between the kitchen refuse and the hard won earnings of the eiders.

1903

A month of dull, dark, unsettled weather, with scarcely any sunshine to speak of, and admitting of but little heliographic communication with the 'shore' … Quite a depressing effect is experienced at such prolonged absence of the land, and the reappearance of the Grampians, though swathed in winter vestments, would be a welcome sight. Our fish supply ceased early in the month and its renewal need not be expected before the month of May. … Our only feathered visitor for the month was a belated bullfinch, who reached us only to die. The eiders and long-tails continue in evidence, and have now the company of four cormorants.

Starfish are always plentiful here, though of course more numerous in summer. All are of the five-rayed variety, including the 'brittle' starfish, which, unlike its fellows, discards its rays on the slightest irritation … In startling contrast was a specimen found in a shallow pool early in the month, and which was quite new to us here. Six inches in diameter, the stranger [*Crossaster pupposus*] appeared all body, with very short rays, of which there were twelve ... the upper surface being richly coloured with concentric rings of crimson.

After nearly nine years on the Bell Rock, Campbell (by then aged 43) concluded his nature notes in April 1904 with characteristic modesty:

Owing to my transference to another station, it now becomes necessary for me to conclude these random jottings. To the patient reader who has cared to follow me through these notes I bid farewell. Written without any pretensions to literary skill or scientific accuracy, they have nevertheless in my case, served to enliven many a weary hour in an isolated calling and have – may I hope? – proved not altogether void of interest to the reader.

His new posting was to the Bass Rock. With its famous colony of gannets, Campbell now spent his spare time observing and ringing these birds, publishing some interesting

Left: John Campbell and his colleagues sharing Christmas on the Bass Rock
Right: The author at the Bell rock in May 2010

and valuable notes (e.g. *Country Life*, May 1914). Indeed, after acquiring a camera, he was the first to photograph a gannet feeding its young one, not an easy subject to secure. To all visitors – his obituary pointed out – 'he was genial and courteous, while to nature photographers he was particularly helpful, as no trouble was too great for him to undertake in order to aid them in their work'.

Latterly, he was transferred as principal keeper, to the Noss Head Lighthouse, Wick, where he continued his observations and studies with his usual keenness. He died in 1916, aged only 55.

seven

This Formidable Work

The great difficulty of landing on [Skerryvore], worn smooth by the continual beat of Atlantic waves which rise with undiminished power from the deep water near it, held out no cheering prospect; and it was not until 1834, when a minute survey of the reef was ordered by the Board, that the idea of commencing this formidable work was seriously considered.

Alan Stevenson, *Encyclopædia Britannica*, 1857

Thomas Smith died in 1815. With Bell Rock lighthouse completed, Robert Stevenson's reputation was now widely recognised so his career, and business, flourished. It would never be challenged again, and his subsequent commissions would now seem somewhat mundane, many of them on the mainland coast. In 1816 he was to replace the old coal beacon on the Isle of May with a handsome stone lighthouse block nearby.

As early as 1801 Robert Stevenson had reported to the NLB how there was a need for a light on the Calf of Man, the southern tip of the island and the approach to many important ports in northern England and Ireland. But neither the NLB nor Trinity House held the appropriate remit. Merchants of Liverpool and the Isle of Man petitioned Trinity House to erect Manx lights but deemed the proposed levies too high. They decided in favour of the NLB since they were cheaper. But doubtless they were also familiar with Stevenson's formidable reputation. So they proposed to the Commissioners of Northern Lights in 1815 – with other representations received from the Chief Magistrate of Greenock and from various trading bodies in the Firth of Clyde – that a light should be erected on the Point of Ayre to make the west coast channel completely safe. The Commissioners then applied to Parliament for power to erect a lighthouse on the Isle of Man and the Bill was passed in July 1815.

But first a loan would be required. The Commissioners were already in debt after the acquisition from the Portland family to the duties from the Isle of May lighthouse; in addition they were also involved at the time in the building of the Bell Rock lighthouse. Lack of sufficient funds slowed progress at the Point of Ayre on Man and a further cause for delay was that the original position of the light had to be altered because the sea was eroding the coast at the rate of 2m (7 ft) per year. But eventually the work commenced late in 1815.

In fact, three years later, Stevenson actually completed three lighthouses on the Isle of Man – one at the Point of Ayre in the north but also two leading lights (one on the hill above the other) on the Calf of Man in the south. The Ayre lighthouse tower was circular, 21 m (69 ft) high and was exhibited sometime between December 1818 and February 1819. A second light called 'The Winkie' was added at Point of Ayre around 1890 but was discontinued in 2010.

Just before his death in 1818, George Dempster, one of the founders of the NLB, had prophesised that before long ' ... the whole northern seas [would be] illuminated like Pall Mall from sunrise to sunset.' Maybe he was overstating the situation and I think he probably meant to say from sunset to sunrise – the hours of darkness. But by this time the NLB were operating 16 major lights and the momentum was still growing.

Sumburgh Head lighthouse, Shetland

In 1821 Shetland was given its first lighthouse at Sumburgh Head, the southernmost tip. The builder was John Reid of Peterhead who was told to make the walls double thickness to keep out the dampness. The most serious offence a lightkeeper could commit was to fall asleep on watch and one of the most notable incidents occurred at Sumburgh in 1871 when two lightkeepers conspired not to report the other for sleeping at his post; one of them was a principal lightkeeper with 23 years' service. Notwithstanding, they were both dismissed.

In 1821, during his annual inspection cruise in the yacht *Regent*, Robert Stevenson visited Eilean Glas lighthouse in Harris. The factor on Scalpay presented them with a live great auk, which some men on St Kilda had caught three years previously. On board was a Fife minister, the Reverend Dr John Fleming (1785–1857) who was also Professor of Natural Sciences in Edinburgh. The two men intended to keep the flightless great auk alive as long as possible and then present its body to his University Museum. Fleming wrote:

The bird was emaciated, and had the appearance of being sickly, but in the course of a few days became sprightly, having been plentifully supplied with fresh fish, and permitted occasionally to sport in the water with a cord fastened to one of its legs to prevent escape. Even in this state of restraint it performed the motions of diving and swimming under water with a rapidity that set all pursuit from a boat at defiance.

However, after Stevenson and Fleming had left the ship, the great auk made its escape near the entrance to the Firth of Clyde. Apparently a short time afterwards a dead great auk was washed ashore near Gourock. A plaque that used to be outside Scalpay lighthouse claimed this was the last of the species in Scotland – in actual fact another was caught in St Kilda in the 1840s. Stevenson's was certainly one of the last of its kind; and before he himself ended his days the great auk as a species would be totally extinct.

Eilean Glas lighthouse, Scalpay, as replaced by Robert Stevenson in 1824 (compare with p79)

Robert soon returned to Eilean Glas, Scalpay, to replace Smith's original lighthouse with a handsome granite tower, 30 m (98 ft) tall and 43 m (141 ft) above sea level. It was automated in 1978, the old Fresnel lens assembly (said to comprise over 300 t of glass) now being on view in the National Museum of Scotland, Edinburgh. It has been replaced by an array of powerful electric lamps, still exhibiting three flashes every 20 seconds visible for 23 nautical miles.

Print of Great Auk by William Macgillivray

The Rinns of Islay lighthouse, on the small tidal Isle of Orsay situated off the south coast of Islay, was completed in 1825. In the 1870s the scientist Lord Kelvin considered it one of the three best revolving lighthouses in the world (along with Buchan Ness and Little Ross on the Solway). The engineer had to use all his ingenuity in seeking new ways to distinguish one light from another. Rinns of Islay was alternately stationary and revolving,

Buchan Ness lighthouse, Aberdeenshire

producing a bright 'flash' of light every 12 seconds, without those intervals of darkness that characterise other lights on the coast. The cost of this lighthouse was between £8,000 and £9,000 and John Gibb of Aberdeen was the contractor for the splendid tower. Six more mainland lighthouses followed, including Tarbat Ness lighthouse, erected in 1826 after 16 vessels were lost in the Moray Firth, and the Mull of Galloway lighthouse built at the most southerly point of Scotland in 1828.

At this juncture it is worth mentioning another Islay lighthouse, but a private one, constructed at Port Ellen, at the south end of the island. It was erected in 1832 by Walter Frederick Campbell (1798–1855), the Laird of Islay, in memory of his wife Lady Eleanor Charteris who died that year, aged only 36. The white-washed, square tower, 17 m (56 ft) tall, was designed by a prominent Scottish architect at the time, David Hamilton who was also responsible for Lady Eleanor's marble casket in Bowmore Church. Above the lighthouse door is carved a lengthy eulogy to her:

Ye who mid storms and tempests stray in danger's midnight hour

Behold where shines this friendly ray and hail its guardian tower.

'Tis but faint emblem of her light, my friend and faithful guide …

Port Ellen was taken over by the NLB in 1924.

Cape Wrath, at the far northwesterly tip of Scotland, was an obvious location for a lighthouse. The name of the headland derives, not from the stormy waters of the area but from the Norse word 'hvarth' meaning 'turning point', for here the Norsemen turned their ships to head home. On the 1814 inspection cruise Sir Walter Scott had been moved to observe:

This dread Cape, so fatal to mariners, is a high promontory whose steep sides go down to the breakers which lash its feet … I saw a pair of large eagles, and if I had had the rifle-gun might have had a shot, for the birds when I first saw them, were perched on a rock within about sixty or seventy yards.

They were of course white-tailed sea eagles which would become extinct in Britain a century later, having suffered from egg collection and persecution from shepherds, gamekeepers and sportsmen through poisoning and shooting – indeed by sportsmen like Sir Walter who did actually have a gun with him!

Cape Wrath lighthouse was built by Robert Stevenson in 1828. The land to the east of Cape Wrath is owned by the Ministry of Defence and has formed part of their military training area since 1933, essentially acting as a safeguard to around 1,000 ha of land which, in 1971, was declared a Site of Special Scientific Interest for its coastal heath and seabird colonies: it is also part of a Geopark for its wonderful geology. The lighthouse was redesignated as a 'rock station' in December 1975 when the lightkeepers' families were relocated to Golspie. On 17 January 1977 a helicopter carried out the Cape Wrath Relief – a history-

Port Ellen lighthouse,
Isle of Islay

Tarbat Ness,
Easter Ross

Left: Cape Wrath lighthouse from the air

Right: Sea Eagle in flight

making moment, as this was the first helicopter relief carried out at a shore-based Scottish lighthouse. The light was automated in March 1998, one of the last to lose its keepers.

Stevenson was also responsible for a third Manx light, constructed at Douglas Head in 1832 for the Isle of Man Harbour Board. This station seems to have ceased operation around 1850; it was then redesigned for the NLB in 1857 by Robert Stevenson's two sons, David and Thomas. It was automated in 1986.

By 1833 Robert was completing Lismore light in the Sound of Mull at the entrance to Loch Linnhe, which leads on to the Caledonian Canal. This is separated from the island of Lismore itself by a Sound ¼ mile broad. Three years earlier the NLB had purchased this 4.5 ha island (Mansedale) for the sum of £500. The light was first exhibited in October 1833 and was fixed white. Robert Selkirk, a descendant of Alexander Selkirk – the inspiration for Robinson Crusoe – was the first principal lightkeeper at Lismore. In 1910 most of the Board's lights were changed to dioptric (incorporating a Fresnel lens to concentrate the light forward) but Lismore and Fidra were left as the only remaining purely catoptric lights in the service (involving simply a parabolic reflector behind the light). In 1940 two lightkeepers at Lismore, under most difficult conditions, rescued two airmen clinging to a piece of wreckage in the sea.

Robert Stevenson's final big challenge was Barra Head, on the island of Berneray at the southernmost tip of the Outer Hebrides. He visited the place on his inspection voyage of 1828 accompanied by his 13-year-old son David who described one of the cottages:

We found several women and a number of children all squatting on the ground round a fire of peats on the floor and without any chimney and except for a few cooking

Lismore lighthouse

Lismore lighthouse in a storm

Barra Head lighthouse on the Isle of Berneray, south of Barra

materials we saw no other articles of furniture whatever. In the other apartment 4 or 5 children of the family made their bed of straw on the floor, with only a single plaid among them ... The women and children had a tolerably clean appearance. The men were more rough. They had neither shoes on their feet nor covering of any kind on their head nor coats on their backs.

The site chosen was 2 km (1¼ miles) from the landing place, on the highest point, Sron an Duin, the name coming from the presence of an Iron Age fort, largely destroyed during the building of the lighthouse, but still impressive nonetheless. There is a reference in the 18th century to the structure once serving as 'a Pharos or watch tower'. Robert Stevenson began construction in 1830 and the following year 48 men were at work. He wrote:

Such is the violence of the wind in this station that the temporary buildings occupied by the artificers were repeatedly unroofed. On the face of the precipitous cliffs the winds and seas acquire a force which the reporter has never experienced elsewhere ... The artificers are often reduced to the necessity of passing the most exposed places on their knees clinging with their hands to the ground ... In one of these storms the lighthouse cart and horse overturned by the force of the wind ... It is the remarkable that the horse sustained no injury.

*Barra Head
lighthouse from
the ancient fort
entrance*

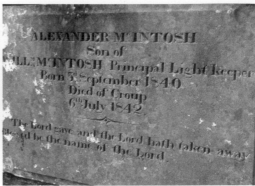

Left: New lighthouse apparatus in Barra Head

Right: Inscription on one of the graveslabs at Barra Head

The tower built of stone is only 18 m (59 ft) high, but the light is about 211 m (693 ft) above sea level and in clear weather had a range of 29 km (18 miles). The total building costs were £12,000 – amongst the highest of any conventional lighthouse built at the time, on account of its remoteness. Three lighthouse keepers lived on the island with their families but their lives were hard. There is a poignant circular little graveyard on top of the cliff where some children of lighthouse keepers were buried, along with a visiting inspector. The presence of the lighthouse maybe contributed to an improvement in circumstances of the local community, providing work, improving access and communications. When Isabella Bird landed in 1863 with Captain Otter in the *Shamrock* there were about 20 local people and 13 lighthouse personnel:

> The whole island was congregated. We received an outrageous welcome … Yet these people were all well dressed, cleanly, and healthy looking, and most anxious to take us to their abodes. We spent some time in one of them, and were regaled with delicious cream, in large, clean, wooden bowls. This was the lightest, cleanest, best appointed Highland hut I ever entered …

Barra Head lighthouse was completed in 1833, its oil burning light converted to incandescent in 1906. Only lighthouse families inhabited the island after the last indigenous people left around 1910; that is until a shore station was established in Oban, 127 km (78 miles) away. A wireless beacon was established there in 1936. On 23 October 1980 the lighthouse was finally converted to automatic operation and the last of the lightkeepers were withdrawn.

Visiting in 1868 the ornithologist HJ Elwes wrote:

> It was the grandest sight I ever experienced, to look out of the lighthouse on a very stormy day, and see oneself hanging, as it were, over the ocean, surrounded on three

Barra Head lighthouse from the sea

sides by a fearful chasm in which the air was so thickly crowded with birds as to produce the appearance of a heavy snowstorm, whilst the cries of these myriads, mingled with the roar of the ocean and the howling of the tremendous gusts of wind coming up from below as if forced through a blast pipe, made it almost impossible to hear a person speak.

Indeed, with its impressive seabird colonies, especially razorbills and most recently four nesting pairs of gannets, I consider this to be my favourite lighthouse.

First Thomas Smith, and now Robert Stevenson effectively designed not just lighthouses but also, of course, the Lighthouse Service. Robert chose to run this along military lines, with uniforms, badges and medals and regular worship – one of the duties of the principal lightkeeper. The aspiring keeper would begin his career as an expectant lightkeeper, soon changed – for obvious reasons! – to supernumerary lighthouse keeper. He gained experience being shunted from one station to another as an extra man, or even as a stand-in for an indisposed or absent senior colleague. Usually within a year or two he had proved his worth until, at the top of the list, he gained a more permanent appointment as an assistant lightkeeper at one particular station, and for a period of five years or so. Thus he received rent-free accommodation for himself and his family, with free fuel but found himself at the bottom of another list. As the years passed he might finally attain the rank of principal lightkeeper with a smart new cap badge – with another to wear on his sleeve – and no need to be lighthouse cook anymore! One of the Board's last job descriptions began:

Not every man is suitable to be a lightkeeper … While primary duty is to keep watch at night, to ensure that his light flashes correctly to character, and to keep a fog watch throughout each 24 hours … a lightkeeper must be a man of parts. He will acquire

129

a good working knowledge of engines … he will know about radio telephones; from his study of the sea he will respect its immense power; he will be a handyman of varying proficiency but mostly of a high standard; he will be a useful cook and a good companion. A lightkeeper will not make a fortune but he will be at peace with himself and the world.

Robert Stevenson brought up his sons with degree of discipline similar to that he expected of his lightkeepers. He wakened the boys at 5 a.m. to begin their working day. Having come from a poor background he saw engineering as an appropriate and prosperous career, but he also encouraged them in other pursuits that he considered worthy.

His eldest son Alan (1807–65) was educated at the High School and the University of Edinburgh where he read Latin, Greek and mathematics with a view to entering the Church. Although not physically strong, with a distinct classical bent, Alan was pressurised by his father to take up engineering – as were David and then Thomas, the youngest. Another son, young Robert (Bob) managed to escape the family tradition by becoming a surgeon in the army. Sadly he was to contract a fatal illness in India and died, at home in Edinburgh, only a year after his father.

It was at the tender age of 12 that Alan had begun his apprenticeship – on an inspection voyage with his father, along with Jane, Bob and their two Smith cousins. In 1824 Alan really began to cut his teeth by assisting his father with the construction of the Rinns of Islay lighthouse and, after winter classes in appropriate engineering subjects, he graduated MA at Edinburgh taking the Fellowes prize in natural philosophy.

On a study tour to Paris, Alan befriended the Fresnel brothers who were making great advances in lighthouse optics. Lighthouses could look quite distinctive in daylight but distinguishing them at night had always presented a problem. Robert had used two adjacent lights or a panel of red glass but now Alan saw the possibilities in Fresnel's flashing lights and lamps revolving behind a screen. This was operated by clockwork, so that winding up the mechanism every 20 minutes or so became a principal duty of lightkeepers and helped keep them awake! For too long Robert had championed reflectors until Alan brought him round to the Fresnel lenses he had seen in France – the dioptric system. Since it was so convenient to Edinburgh, the Forth island of Inchkeith was often used to test new developments so, in 1835 it became the first lighthouse in Britain to employ such revolving lenses, and the Isle of May lighthouse had the first fixed light with lenses the following year.

David Stevenson later described the innovation:

When Inchkeith was changed from a fixed to a revolving light, a certain old lady [ashore in Granton], who had beguiled many a sleepless hour in watching it, was greatly puzzled by its successive appearance and eclipse, and declared in the morning that the poor lightkeeper was much to be pitied, for 'no sooner was his lamp lighted than it went out, and if it had been lighted once, it had been lighted a hundred times!'

Inchkeith lighthouse, Firth of Forth

In 1828 Alan had compiled a gazetteer on *The British Pharos*, the first ever almanac for seamen. He was by now effectively a partner in the family firm – though nothing was formalised until 1833. With his father Robert now heavily engaged in designing bridges, harbours and river works as well as lighthouses, Alan took a bigger role for the NLB, for whom he was appointed Clerk of Works in 1830 – a post his father had held 25 years earlier. It was around this time that the NLB moved into 84 George Street in Edinburgh, where it remains to this day. Robert Stevenson moved into the floor above them until he finally retired from the NLB in 1843, at the age of 70. Alan assumed the post of chief engineer for the NLB but in effect found that he was still functioning as their Clerk of Works as well. Robert then resigned from the family firm on the death of his wife Jane in 1846. His youngest son Thomas was now made a partner alongside elder brothers Alan and David. Robert died on 12 July 1850, Alan saying of him: 'A high sense of duty pervaded his whole life and he died calmly … ' He is buried in New Calton Cemetery in Edinburgh.

The Commissioners placed a marble bust of Robert in the library room of the Bell Rock (but now in their George Street headquarters): 'In testimony of the sense entertained by the Commissioners of his distinguished talent and indefatigable zeal in the erection of the Bell Rock Lighthouse.'

Although Thomas's only child, Robert Louis was born a few months after the death of his grandfather it is obvious that he came to understand him well:

… an intrepid old man, who, when told he was dying, fretted, not at approaching death, which he had faced many a time, but at the knowledge that he had looked his

Left: Headquarters at 84 George Street, Edinburgh
Right: Bust of Robert Stevenson from Bell Rock, now in NLB HQ George Street, Edinburgh

last on Sumburgh, and the wild crags of Skye, and the Sound of Mull, that he was never again to hear the surf break on Clashcarnock, never again to see lighthouse after lighthouse open in the hour of dusk their flowers of fire or topaz and the ruby interchange on the summit of the Bell Rock.

While Robert's career and reputation had been defined for posterity by the Bell Rock, Alan's would be by Skerryvore. This is a 13 km long reef, 18 km (8 miles) south of Tiree. Between 1790 and 1844 over 30 ships had foundered there. In *A Star for Seamen* Craig Mair commented: 'Like the Bell Rock, this was a much-dreaded reef: if ships did not crash into it, they sank elsewhere trying to avoid it.'

Robert Stevenson had landed upon Skerryvore in 1804, taking every opportunity to view it thereafter. In 1814 he was accompanied not only by Commissioners, but also by his friend and companion, Sir Walter Scott. Stevenson was determined to convince them of the need for a lighthouse:

Loud remonstrances on the part of the Commissioners, who, one and all, declare they will subscribe to his opinion, whatever it may be, rather than continue the infernal buffeting.

Skerryvore lighthouse
(Iain Thornber)

Not surprisingly the landing was tricky. Typically Scott was somewhat distracted:

> I saw nothing remarkable in my way excepting several seals, which we might have shot, but, in the doubtful circumstances of the landing, we did not care to bring guns … The rock was carefully measured by Mr S. It will be a most desolate position for a lighthouse, the Bell Rock and Eddystone a joke to it.

It was not until 1834 that the NLB took the decision to build on Skerryvore and, by then getting on in years, Robert was to entrust this formidable work to his 27-year-old son Alan. Robert had been eight years older than this when before, with some reluctance, the NLB had entrusted him to take on the Bell Rock. Alan – in his narrative of the building of Skerryvore – reflected on the challenge:

> From the great difficulty of access to the inhospitable rock of Skerryvore, which is exposed to the full fury of the Atlantic, and is surrounded by an almost perpetual surf, the erection of a Light on its small and rugged surface has always been regarded as an undertaking of the most formidable kind.

The sailing vessel *Pharos,* veteran of Bell Rock days, was again brought into service until a steam tender was ready. Again, like Eddystone and Bell Rock, the situation demanded a stone tower of interlocking blocks (some 2.5 t each in weight) but the reef was more extensive, difficult to approach and lay open to the full fetch of the Atlantic Ocean. Finally, although a similar distance from the nearest land, the land in question was not the mainland (which is 80 km or 50 miles away) but an island in the Inner Hebrides – Tiree. A channel had to be found in the reef to facilitate access, and negotiating the slippery, polished rock was likened to climbing the neck of a bottle. An elaborate shore station and harbour was established at Hynish in Tiree with stone for the tower being sourced from a quarry on Earraid in Mull, between Tiree and the mainland. The accommodation barrack out on the rock did not survive its first winter so had to be rebuilt. The foundation pit sunk into hard gneiss rock necessitated the use of dynamite. Considering the confines of the reef it is a wonder that no one was seriously hurt. The whole project would take seven years – most of Alan's career – and (including the Hynish facility with its keepers' family accommodation) cost as much as £86,977.17s.7d.

Alan had proposed a taller tower than Bell Rock, 48 m (158 ft) high with an enormous flared base 14 m (46 ft) in diameter. Altogether 4,300 t of granite were employed, twice as much as Bell Rock and 4.5 times as much as Smeaton's Eddystone. Furthermore Alan had taken the bold decision to save time and money by only interlocking the blocks horizontally, relying on the weight of the structure and mortar to bind each of storeys on the vertical plane. This resulted in the structure yielding somewhat to the pounding of the waves, causing one lighthouse keeper many years later to observe how the Bell Rock would shudder in storms whereas Skerryvore would not. His younger brother Thomas, fast developing a skill with

The stone quarry on Earraid, Isle of Mull

Fresnel lenses, helped install the lantern – Skerryvore's 'crown of glittering glass'. Alan had incorporated eight Fresnel lenses with reflecting prisms below and mirrored reflectors above to maximise the beam's intensity. Skerryvore was finally exhibited on 1 February 1844. The light could be seen from Barra, around 60 km (38 miles) to the west, and allegedly from the summit of Ben Nevis, 145 km (90 miles) east. The resulting tower, Robert Louis Stevenson judged, was: 'The noblest of all extant deep-sea lights.'

In 1884 Robert Louis Stevenson went on to call his Bournemouth house 'Skerryvore' and had the following written above the door:

For the love of lovely words, and for the sake
Of those, my kinsmen and my countrymen,
Who early and late in the windy ocean toiled
To plant a star for seamen, where was then
The surfy haunt of seals and cormorants:
I, on the lintel of this cot, inscribe
The name of a strong tower.

In 2006 the author Christopher Nicholson believed that there is not a tower in the world that can surpass Skerryvore in its beauty of design. At 48 m (158 ft) it remains the tallest

The Skerryvore shore station, signal tower and harbour at Hynish, Isle of Tiree

lighthouse in Scotland with 151 steps. The tallest in the UK is Bishop Rock lighthouse, redesigned by William Tregarthen Douglass, son of Sir James who had built the current Eddystone lighthouse. It is 49 m (161 ft) tall, having lost two metres when a helipad was established on its roof.

Skerryvore was one of the last Scottish lighthouses to use rectangular glass panes. From now on Alan employed the now familiar diagonal glazing around lighthouse lanterns. At the same time England introduced lozenge-shaped panes of glass but Alan's were triangular. They gave added strength against the wind but more importantly they impeded the intensity of the beam less and avoided the impression of a flashing light as the beam passed between vertical glazing bars. They were first installed at Noss Head in Caithness, and then Ardnamurchan, both in 1849, and they are now standard for NLB lighthouses.

Communication between rock lighthouses to the shore was initially by visual means only. The Signal Tower at Hynish served to transmit and receive semaphore signals to or from Skerryvore lighthouse (visibility permitting) at pre-arranged times, or otherwise in the case of an emergency. A similar signalling system had been in operation between Bell Rock and its shore station in Arbroath (now a museum); both shore stations are now museums. Hynish had limitations as a shore station for the Skerryvore families, so in 1892 was abandoned in favour of Earraid on the Island of Mull with the Dubh Artach families (ultimately both relocated to Oban). On Earraid an observatory tower came to be built to view both lighthouses on the distant horizon.

In mid-July 1940 German planes attempted to bomb Skerryvore, one exploding near the tower, cracking two lantern panes and shattering an incandescent mantle. However, on the night of 16 March 1954, fire broke out in the tower – every lightkeeper's nightmare – and it quickly spread. Despite the keepers' heroic efforts, they were ultimately forced to abandon the blazing lighthouse altogether and seek refuge on the bare rock outside. It is said that a guillemot climbed on to the rock at night and settled beside the men until dawn. Fortunately it was an unseasonably fair and calm night. The next day was the scheduled relief so the *Hesperus* turned up and was able to rescue the three men. The damage proved extensive so an unmanned lightship was deployed until a temporary Dalen light was established in the tower. This was a Swedish invention, involving an automatic light powered by bottled acetylene. Repairs to the lighthouse took four years to complete, with a new, generator-powered electric light installed; but the successful operation of the Dalen light helped convince the NLB that automation of major offshore lights was feasible.

In the 1970s, another accident occurred at Skerryvore when the relief helicopter was washed clean off the helipad by a freak wave; fortunately no one was in the helicopter, but it had to be written off.

Skerryvore was the last rock lighthouse in Scotland to become automatic in 1994, while in October 2005 it was the last Scottish foghorn to be switched off. After experimenting with bells, gunfire, horns and gongs, steam or compressed air units eventually became standard. The first to be installed by the NLB – as late as 1875 – was at St Abb's in Berwickshire. Once, in 1899, the Inchkeith foghorn sounded more or less continuously for 130 hours, much to the dismay and irritation of the folk living on the Fife coast! The universal application of electronic navigation systems and radar eventually rendered foghorns redundant: the first such to be operated were at Cape Wrath, Copinsay, Fair Isle North and Rattray Head in 2001.

With Skerryvore Alan Stevenson had risen to a challenge that some say exceeded that of his father. In 1843 he went on to build Little Ross lighthouse in Kirkcudbright, which became notorious in 1960 when one lightkeeper murdered his companion: the station was automated almost immediately afterwards! In 1844 Alan also constructed the Low Light on the Isle of May, but this became redundant in 1887 when the North Carr lightship became operational.

Alan married Margaret Scott in 1844 with whom he had a son – Robert Alan Mowbray (1847–1900) an art critic – and three daughters, one of whom became an authoress. By now Alan was a member of the Institution of Civil Engineers, and a fellow and council member of the Royal Society of Edinburgh. He was a remarkable engineer, receiving acclaim both at home and abroad: he was invited to design a lighthouse for Singapore for instance. In 1840 he received an honorary doctorate from Glasgow University, along with various medals from the kings and emperors of Russia, Prussia and the Netherlands recognising his merit as a lighthouse engineer.

With 30 lighthouses now on the NLB books, Alan was kept exceedingly busy. In May 1844 for example, he left Edinburgh for Inverness, a coach journey of 20 hours. Next morning he visited Cromarty and Fortrose, then Chanonry, from where he was rowed to

Chanonry Point lighthouse at Rosemarkie, Easter Ross

Inverness, continued to spend the Sabbath (a day of rest) in Forres before visiting Covesea Skerries, then Fraserburgh via Elgin and Banff. He inspected Kinnaird lighthouse then Peterhead harbour, Buchan Ness and Girdleness before returning home via Dundee – all in the space of eight days. This was the first of several such trips that year including the annual inspection voyage in July and a trip to the Isle of Man where he received a ducking at the Calf of Man (and then another at Cape Wrath a few days later). He had never enjoyed a robust constitution and soon began to suffer from lumbago.

His punishing routine and poor health probably resulted in his then undertaking a succession of relatively straightforward mainland lights (Covesea, Chanonry and Noss Head for example). Undaunted though, in 1849 Alan embarked upon Ardnamurchan lighthouse on the most westerly point of the British mainland (6°13′W). During the three years it took to complete, scurvy broke out among the workmen. The tower, with a flavour of Egyptian style decoration, is 35 m (118ft) high with 152 steps inside.

Alan then went on to tackle Sanda (in 1850) situated off the south end of Kintyre. Since the *Christina* of Glasgow was lost with all hands in 1825, there had been a demand for a light on this island, which forms the turning into the Clyde from the north Channel between Scotland and Ireland. Trinity House proposed moving the Mull of Kintyre light to Sanda, but the Commissioners deferred. Further wrecks finally necessitated building a separate lighthouse on the summit of Ship Rock, a detached crag at the south end of Sanda. Alan set two novel stone towers against the face of the rock, enclosing the access from the accommodation block below to the whitewashed light tower on top, a design still unique in Scotland. This resulted in an NLB overspend for 1848. It was first lit in 1850 but did not entirely stop ships coming to grief. On 19 October 1970, the Dutch cattle ship *Hereford Express* went aground. The livestock that survived the wreck had to be destroyed anyway by SSPCA officers who flew out to Sanda by helicopter.

None of Alan's lights subsequent to Skerryvore were to prove particularly challenging – perhaps deliberately considering his poor health – but were nonetheless demanding of his

Left: Ardnamurchan lighthouse, Inverness-shire from the air
Right: Ardnamurchan lighthouse and foghorn

time and energies. With increasing work for both the family firm, and for the NLB, Alan took on Alan Brebner as an assistant: he was the son of the man Robert Stevenson had employed as mason on Bell Rock, and who would rise to become a partner in the firm. Under Alan Stevenson in 1851, the firm went on to build both Hoy High and Hoy Low as leading lights for the famous naval anchorage of Scapa Flow in Orkney, followed by Arnish in Lewis in 1853.

Arnish, at the entrance to Stornoway harbour, is an interesting case for it is here that Thomas Stevenson was able to test his latest innovation – the 'apparent light'. Seamen had long lamented that many seamarks were unlit – both 'blind' and 'dumb' they said. Thomas

Sanda lighthouse, Kintyre (Rick Price)

came up with the idea of using the beam from a nearby lighthouse to bounce off glass prisms mounted on the seamark. The deception was to prove so perfect that 'fishermen will not believe that there is not a light there.' Arnish was chosen as the first site to showcase this device, which was to remain in operation for fifty years. Finally worn out, it had to be replaced in 1905 but this time by a gas light.

In 1851 Alan invited a group of eminent astronomers on the annual lighthouse tour of inspection and diverted the ship north of Bergen so they could view the solar eclipse. In the party was the Astronomer Royal for Scotland, Charles Piazzi Smyth, who would go on to make measurements of the pyramids of Giza and advance controversial theories on their astronomical and astrological significance. Alan always retained a love of the arts; fluent in Latin, Greek, French, Spanish and Italian, and he translated and memorised Greek poetry. Indeed it is a characteristic of his lighthouse architecture that he incorporated discrete Classical and Egyptian influences.

In 1853 Alan had to retire prematurely due to ill health – Bella Bathurst suspects he might have had multiple sclerosis – and, with no pension, he died 'of general paralysis' in somewhat reduced circumstances in 1865 to be buried near his father. He had pushed himself to his absolute limits and was responsible for a dozen major lighthouses in his short ten-year career as NLB Chief Engineer, as against 20 constructed by his father in his 44

Arnish lighthouse, Isle of Lewis. Thomas Stevenson's 'apparent light' was where the buoy now is

years. Skerryvore will always be his memorial, his obituary in the *Journal of Civil Engineers* summing up:

> … in personally conducting that great work, during a period of five working seasons, his courage and patience were severely tried, and his abilities as an Engineer were fully tested. They were found equal to the task of successfully accomplishing what will ever be regarded as a triumph of lighthouse engineering, and as perhaps the finest combination of mass with elegance to be met with in architectural engineering.

The Commissioners added their own unusually heartfelt tribute, noting: '… their deep and abiding regret for the loss of a man … whose genuine piety, kind heart and high intellect made him beloved.'

eight

Sources of Safety

Among the many works of man which prove the truth of the saying that 'knowledge is power', we must not omit these solitary towers, often half-buried in the surge, that convert hidden dangers into sources of safety, so that the sailor now steers for those very rocks which he formerly dreaded and took so much care to avoid.

Thomas Stevenson, *Lighthouse Construction and Illumination*, 1881

During Alan's tenure with the NLB his younger brother David Stevenson (1815–86) had been building a reputation for the firm in river engineering, while Thomas (1818–87), the youngest, had particular skills in optics, invention and science. It was he who invented the meteorologist's Stevenson's Screen. At David's own insistence the brothers were appointed jointly as consultant engineers with the Fishery Board in 1851 and, on Alan's retirement, David took over his role as NLB Chief Engineer. He soon convinced them to make this also a joint position with Thomas. Nine new lighthouses were already on the drawing board, dozens of beacons and buoys with the mountain of associated paperwork. Despite poor health, David presided over what are termed the 'golden years' of lighthouse construction in Scotland. David and his wife Elisabeth had eight children, two baby sons did not survive but the other two – David Alan and Charles Alexander – became lighthouse builders.

David's brother Thomas was the youngest of Robert's sons to enter the lighthouse trade and despite his many accomplishments, will probably be best remembered as the father of the author Robert Louis Stevenson. Initially Robert Louis showed little interest in the family business but, bowing to pressure from his father, he subsequently came around to make significant contributions in the field. Thomas assumed the major role in the final installation of the Dubh Artach light, a tricky project reminiscent of the Bell Rock. After his brother David retired, he took over the operation of the family business. Thomas was a Calvinist, and a firm adherent of the Christian faith. In 1848 he married Margaret Isabella Balfour (whose brother was an apprentice engineer in the firm) and two years later, just months after his own father Robert Stevenson had died, Thomas's only child Robert Louis was born.

The two brothers, David and Thomas, made a fine team but were as different as chalk and cheese. David was small, bespectacled, dapper, popular, reliable and addicted to work. He was the better engineer and, as the capable managing partner in the family business, he built on the reputation it had established under his father and elder brother. Thomas, on the other hand, was tall, quiet, stern, melancholic, with rather strange and unpopular views but nonetheless able and dedicated when necessary. Much to his father's annoyance, he was at first more interested in book collecting, literature, writing and natural history than engineering. Robert once chastised Thomas when he found some manuscript pages of fiction in a drawer: he was urged to 'give up such nonsense and mind your business.' It is strange then that Thomas would prove so exasperated with his own rebellious son's literary ambitions, rebuking Robert Louis that writing was 'no profession'. In later life father and son would later enjoy corresponding with one another over literary matters and even correct each other's writings and speeches. Bella Bathurst perceptively summed up the complexities of Thomas's character – 'a man forever fighting his own contradictions'.

The earliest beacons in Britain had burnt solid fuel, mostly coal. Around 1750, 21 of the 25 navigation lights in Britain, nearly all of them in England, were still coal-fired. This was a convenient fuel, burning longer than wood in wet weather. They were however difficult to control, demanded constant and laborious attendance, but worst of all they could easily be confused with domestic or industrial fires. It was Thomas Smith's reflecting oil lamps that finally replaced the coal beacon on Little Cumbrae in 1793.

The Isle of May lighthouse today

When the new Isle of May lighthouse and its oil lamps were lit in 1816, the coal fire nearby was extinguished, after 181 years' service. The only surviving cast iron grate, chauffer or cresset is from St Agnes in Scilly, redundant since 1809 when Peninnis Head was built nearby, and is now to be found in the Abbey Gardens on Tresco. The last coal beacon in England was not replaced in 1823.

Smeaton's Eddystone lighthouse had been lit by a few 'miserable tallow candles' with mirror reflectors for the first 40 years of its service. By the early 1800s Arctic whaling was a major industry from ports such as Dundee, Peterhead and the Northern Isles. The first new Argand lamp, using silver-plated copper reflectors, and fuelled with sperm oil, was installed in Inchkeith in 1804. Thereafter, crude whale oil or finer spermaceti was for a long time the main illuminant for oil lamps in most Scottish lighthouses. It had a tendency to become viscous in cold weather though, so, gradually this rather costly and sooty commodity came to be replaced by cleaner colza (rapeseed) oil. Whale oil lamps needed modification for colza use, but the quality and erratic supply of colza oil, not to mention its increasing cost, proved disadvantageous. Other oils – olive, seal, even herring – were tried until, by 1847, shale oil from the Lothians became readily available, to be trialled in Girdleness lighthouse, and came to be first installed shortly afterwards at Pentland Skerries and Pladda.

Girdleness split level lighthouse, Aberdeen

From mineral oil, paraffin could be extracted and, gradually, as its price reduced, oil lamps came to be replaced by bright mantles burning paraffin vapourised under pressure. The price dropped from 1s 4d per gallon in 1870 (when only half the price of colza) to 4d by 1894. With 76,000 gallons a year being consumed around Scotland's coasts, there was a saving to the NLB of £10,000. By 1870 David was famously promoting the use of paraffin instead of the expensive colza oil. In 1872 Dubh Artach became the first rock light

Left: Davaar island off Campbelltown, Kintyre
Right: Davaar lighthouse

to utilise paraffin. As the better engineer, in some ways it is perhaps unfortunate that David's individual achievements are always considered in conjunction with those of his still-talented partner and brother.

By the mid-19th century British trade was expanding and steamships were becoming commoner. The Board of Trade, the government body handling all things mercantile, wished to rationalise lighthouse administration in Scotland, England and Ireland. The Merchant Shipping Act of 1853 accorded Trinity House more power over the NLB and the Irish Board. Regular wreck returns were initiated at this time too revealing how, between 1859 and 1866 there were an average of 24 vessels coming to grief every year around the Scottish coast. When in 1853 David Stevenson succeeded his brother Alan as NLB Engineer he was already engaged in building a light at Davaar near Campbeltown. He was also busy on a new tower on North Ronaldsay in Orkney; this would re-commission the light that had been installed there by Thomas Smith in 1789, to be discontinued in 1809 once Start Point had been built on Sanday. North Ronaldsay's red brick tower was given two distinctive white bands in 1889.

North Ronaldsay,
Orkney, with sheep
(Ken Wilkie)

A third lighthouse was also in progress – on Out Skerries, a group of islands at the easternmost extremity of Shetland. Two of the islands are inhabited and joined nowadays by a bridge. The first light on Out Skerries was a temporary structure built on the third, uninhabited island of Grunay in 1854. It was here that the NLB wanted to build the permanent lighthouse and where the keepers were housed, but this site was over-ruled by the Board of Trade and Trinity House. In the end the tower was erected on a skerry two miles further east at a cost of £21,000, vastly more expensive than the Stevensons' estimate. Thirty metres high (98 ft), it was completed four years later.

Muckle Flugga, Shetland

Out Skerries fitted well with the Government's plans because in 1854 Britain had become involved in the Crimean War, which was to last three years. Although far from the actual theatre of conflict, the British Government set about blockading Russian supply ships to and from the Arctic, which in turn highlighted the urgent need for additional strategic lighthouses in Orkney and Shetland. To date there were five in Orkney but the only lighthouse in Shetland was still Sumburgh Head, built by Robert Stevenson in 1821. In only one year (1854) David and Thomas were to establish two more, Out Skerries and – the most challenging assignments yet – North Unst or, as it is now known, Muckle Flugga.

A jagged outcrop at the most northern point in the British Isles, Muckle Flugga was right in the path of North Atlantic storms. In 1850 Alan Stevenson had – with some difficulty – leapt ashore to snatch a rock specimen. Ever since then the NLB had it in mind to establish a lighthouse somewhere in the vicinity. In February 1854 David Stevenson was despatched

Muckle Flugga, Shetland

there to investigate and, unable to land, he deemed it impracticable to attempt anything. He was '… adamant that the seas around the Shetland coast made building a lighthouse in the area impossible, impractical, dangerous, too expensive, and any ship that took that route was mad anyway.'

Indeed, based on his conclusions, on the losses of the barrack on Skerryvore and on the recent destruction of Nicholas Douglass's iron lighthouse at Bishop Rock in the Scillies, the Commissioners reported back to the Board of Trade how it would be: '… culpable recklessness as regards the lives of the lightkeepers to erect even a temporary lighthouse on the Flugga.'

David Stevenson favoured one of three other possibilities, especially Lamba Ness to the northeast. Trinity House persisted however and when their team managed to land without any difficulty on an unbelievably calm June day, they convinced both themselves and the Board of Trade. Stevenson reluctantly capitulated and in July his brother Thomas and Alan Brebner made the difficult climb to the summit of the Flugga and found a flat shelf that could accommodate at least a temporary light. Work had to begin immediately. David considered:

> It was no joke to have the responsibility of having lights ready for our Fleet after the Government had given their opinion that these lights were necessary and should be exhibited before winter sets in … the whole of the materials and stores consisting of water, cement, lime, coal, iron work, glass, provisions, etc, and weighing upwards of 100 tons had to be landed at an exposed rock, and carried up to the top on the backs of the labourers.

Twenty masons and labourers lived in the temporary iron barracks on the rock whilst others were rowed daily from the shore base in Burrafirth. Incredibly it took 26 days to erect the temporary light, which was first exhibited on 11 October 1854. A foreman Charles Barclay – who had lost a hand working at Skerryvore yet a veteran also of Barra Head – stayed overwinter to supervise cutting steps and maintain the light. He reported 'sheets of white water' regularly driven 15 m (50 ft) over the lantern.

On one December morning: ' … the sea was all like smoke as far as we could see and the noise which the wind made on the roof of our house and on the tower was like thunder.' The massive iron door of their dwelling smashed open and admitted a metre high wall of water which swirled around inside before retreating along with everything that that had not been tied down. Then another and a third before the men could secure the door again: 'We have not a dry part to sit down in or a dry bed to rest in.'

All this at nearly 70 m (230 ft) above sea level, higher than Nelson's Column in Trafalgar Square. Over the winter the NLB's relationships between the Board of Trade and Trinity House deteriorated to the point of the Commissioners offering their resignation, only to be told: 'While the buildings at Muckle Flugga could not be pronounced free from risk, that risk would not be greater or probably so great as would be incurred by shipping if the lighthouse were removed.'

Orders to proceed with the work were given in June 1855 so, in desperation, and knowing that everything had to be man-handled up the rock, David Stevenson was forced to adopt 'an untried experiment in marine engineering' – he would build his 20 m (66 ft) tower with bricks. Its foundations were sunk 3 m (10 ft) into the rock and the walls were over 1 m (3 ft) thick. A circular protective wall enclosed the lighthouse yet at times in the future its keepers would have to progress around the courtyard on their hands and knees, and occasionally found live fish thrown over the wall! Over 100 men worked on the site and a permanent light was exhibited on 1 January 1858.

In spite of all possible economies, Muckle Flugga cost £32,000. Against many storms of seismic fury the structure has stood firm, the sea never penetrating the tower and the light never having failed. At 61°59′N with only 5 hours 8 minutes operation required in midsummer – 'simmer dim' as the Shetlanders call it – the light burnt one gallon of paraffin per night; in midwinter on the other hand three gallons were needed over 18 hours 11 minutes. It is interesting to note that on 18 June 1869 a 19-year-old Robert Louis Stevenson visited Muckle Flugga with his father, Thomas, Engineer to the Board:

> … the most northerly dwelling house in Her Majesty's dominion … We pulled into the creek between the lighthouse and the other rock … making a leap between the swells at a rusted ladder laid slant-wise against the raking side of the ridge. Before us a flight of stone steps leading up the 200 feet to the lighthouse in its high yard-walls across whose feet the sea had cast a boulder weighing 20 tons.

To supplement food supplies the keepers collected birds' eggs in the summer, while one even tried to keep hens (needless to say this experiment was not a success). The supply boats from the shore station at Burrafirth had their work cut out, especially in winter when relief might be delayed for 20 days or more. In summer the annual trip with the four families from shore always proved popular. One intrepid keeper even managed to swim all round the rock – in instalments determined by the swell – and landed on every skerry. He also used two upright poles with a crossbar to have himself a game of football on Britain's most northerly pitch – needless to say he always won. Another keeper kept fit by running up and down the 71 steps in the tower, and any others he could find. In 1968/9 a new dwelling block was built in space saved by electrification: this replaced the primitive conditions where, as one principal put it, lightkeepers slept in a crow's nest and ate in a cell. The advent of helicopters greatly facilitated the relief operations until the lighthouse was automated in 1995.

Surprisingly, up to the middle of the 19th century Skye had not figured in any plans to illuminate Scotland's shores, but the advent over sail of more manoeuvrable steamships opened up the Inner Sound east of Skye as a safe and sheltered sea route. As long ago as the 13th century, after his defeat at Largs, King Haakon of Norway had navigated this passage (hence the name Kyleakin) while a local legend states that a Norwegian princess known as 'Saucy Mary' used to hang a chain across the narrows from Castle Maol and charge a toll to allow boats through.

Left: Kyleakin lighthouse on Eilean Ban before the bridge
Right: Kyleakin light passed by sailboat Britannia

Thus lighthouses were proposed to mark this passage – including Kyleakin, Isle Ornsay and South Rona. All three were constructed by David and Thomas Stevenson and first lit on 10 November 1857. They bear the familiar lines of a Stevenson lighthouse but both Ornsay and Kyleakin came to incorporate an ingenious innovation of Thomas's. With the famous optical firm of Chance Brothers in Birmingham, he had developed a 'condensing' system of prisms that reduced the intensity of the beam in some directions whilst optimising it in others, ideal for narrow channels. From the handsome 19 m (62 ft) tower of Isle Ornsay for instance, the light could shine brightest down the Sound of Sleat, slightly less so northwards to Kylerhea, and dimmest of all directly across the narrows from the lighthouse itself.

Notwithstanding his 'apparent light' at Arnish, perhaps Thomas Stevenson's greatest contribution to lighthouse optics was his development of the holophotal system. This ingenious device combined the whole sphere of rays diverging from a light source into a single beam of parallel rays. It was first adopted on a large scale at North Ronaldsay in 1851 and thereafter became universal throughout the service.

On South Rona, at the northern end of Raasay on the Inner Sound, a widow named Janet Mackenzie had maintained a light in her window to enable local fishing boats to avoid dangerous rocks at the harbour entrance. According to Captain Henry Otter, of HMS *Comet*:

> For 10 years she has kept this light burning except in light summer nights, and in stormy weather when vessels are seen beating about, she puts up two ... Many fishing boats owe their safety from the storm to the poor widow's lights when beating up the Sound of Raasay in long winter nights and, unable to contend against the terrific squalls that blow from the Skye shore, they anxiously watch for a glimpse of the narrow belt of light ... The only assistance she has ever received was £20 some years

Left: South Rona, off Raasay, Skye
Right: Isle Ornsay, Sound of Sleat

ago from the Trinity Board. Alan Stevenson advised to reward her 'praiseworthy exertions' with financial aid so the Commissioners accordingly sent a further £20 to her, asking Captain Otter to convey the gift.

One lighthouse keeper, David Dunnet, the first of four generations in the service, was principal at South Rona for 16 years – 1878–95 – quite a tour of duty. All three lighthouses built along the Inner Sound had accommodation for three lighthouse keepers and their families. The lightkeepers at Kyleakin had always been convinced that the house on Eilean Ban was haunted, one nonchalantly admitting how he soon got used to the voices and odd noises! When Kyleakin and Ornsay were converted to gas in 1898 only one keeper was required. Like many early lighthouses sperm oil fuelled the light first, then a paraffin vapour system, until in 1960 it was converted to acetylene gas.

Kyleakin was fully automated in 1960 followed two years later by Ornsay. In 1963 both redundant accommodation blocks were sold to Gavin Maxwell (1914–69). The author of *Ring of Bright Water*, Maxwell had lived on the mainland shore opposite until his house burnt down with one of his tame otters inside. He planned to rent out the accommodation block on Ornsay but in the end was forced to sell it again. On Eilean Ban, Maxwell intended establishing a small animal park, with an eider 'farm' on a neighbouring island to harvest the valuable down. He only lived at Eilean Ban for a year or so until his death from cancer in 1969. The light was finally decommissioned in 1993 when the Skye Bridge opened above it. The Eilean Ban Wildlife Trust, who now own the island, has restored the house and tours (including the redundant lighthouse) can be arranged. David A and Charles Stevenson were also to build other small beacons around Skye, with a final minor light on Sleat Point being constructed in 1933–4.

Cantick Head lighthouse (1858) in Orkney was built to mark the southern entrance of the naval base in Scapa Flow, while Bressay lighthouse (1858) guides shipping into the safety of Lerwick harbour from the island, which gives the light its name.

The tense triangle between the Board of Trade, Trinity House and the Commissioners continued however. The Board prevaricated on Skeirvule in the Sound of Jura, which remained unlit for five years until they approved the estimates. It was one of the first lights to be automated in 1945. They also argued for months as to whether the Butt of Lewis should show a fixed or a flashing light: Ruvaal on the Sound of Islay was to suffer similarly. The NLB were over-ruled as to the site, height and the light signature at Rubha nan Gall in the Sound of Mull. Nearly 200 letter pages were exchanged before the light was finally shown in 1857. This station, within walking distance of Tobermory, was reduced to one-man operation in 1898 and became another of the early lights to be automated in 1960.

Trinity House's elder brethren visited the west coast in 1857 and two years later a Royal Commission cruised round Britain and Ireland to report on lighthouses. One of its members, Captain (later Admiral) Sir James Sullivan was particularly damning. Still smarting from the Muckle Flugga confrontation, he criticised the ignorant landsmen of the NLB for favouring cheaper inshore sites instead of the 'outer rocks'. Yet he was happy to contradict himself – after praising the Stevensons' buildings as 'very beautiful and the work never wants repair' he went on to complain how they were 'more like gentleman's houses inside than lightkeepers' houses need be.'

Kyleakin lighthouse with Skye Bridge and Cuillin

Skeirvule lighthouse, off Jura

Rubha nan Gall lighthouse, Tobermory, Isle of Mull, looking to Kilchoan, Ardnamurchan

Ruvaal lighthouse, Isle of Islay

Optic from Ruvaal light after automation, now in the gardens of Colonsay House

Not content with that, Sullivan immediately condemned the Ruvaal accommodation as 'little better than dog kennels than anything else' and was astonished that any human being could be got to live in them. The keepers that did reside in them observed how the light tower would oscillate in high winds, due in truth to Board of Trade false economies, an attitude the Royal Commission noted as keeping 'economy rather than progress in view.' They identified how the Scottish and Irish Boards had two masters, and the Board of Trade had even once over-ruled Trinity House. In the end the Commission concluded:

As matters now stand, the whole management of the lighthouse service appears to be impeded by the opposing action of three separate governing bodies, and it does not clearly appear to what advantage is gained to counterbalance the delay which results from this complicated systems ... There can be no doubt that of all the British lighthouses visited, the Scotch are in the best state of general efficiency, the English next, and the Irish third.

This merely reiterated one keeper's observation: 'If it was Stevenson-built, it was built to last.'

In Ireland, lighthouse construction was slow to gain any momentum. Although, as long ago as 1665, Charles II granted permission for six lighthouses to be built on the Irish coast (including a proper tower at Hook Head), the venture was not a success. In 1704 Queen Anne transferred the management of all (three!) Irish lighthouses to a group of men known only as 'Commissioners', principally the Customs and Harbour Board of Dublin. In 1786, the Commissioners of Irish Lights as such was created in the same act as the Commissioners of Northern Lights. However, still with only 14 lighthouses and the system in considerable disarray, the Commissioners in Dublin – by then the second largest port in the British Isles seem to have been responsible for all Irish lights. Later, when the Republic of Ireland was created, it was decided to retain just one body (CIL) to control all the island's lights, with Northern Ireland having appropriate representation on the Board.

Where Scotland had readily adopted the lens or dioptric system the old catoptric reflectors were still predominant in England and Ireland. The Scots were ahead on the scientific side too, the NLB engineers studying optics and introducing many innovations. Workmanship was excellent while liaison between engineer and the optician who would assemble the lighting apparatus (each specifically tailored to its geographical setting) were accompanied by large-scale drawings that helped eliminate mistakes. The powerful influence of Robert Stevenson and his sons was obviously being acknowledged. Even the staff at the coal-face did not go unnoticed: 'The Northern Lightkeepers were certainly an intelligent and respectable class of men, for their station in life ... '

By now it was noted how Scotland had one lighthouse for every 39.5 miles of coast, Ireland one for 34.5 miles and England one for 14 miles. (France to whom they had favourably compared Scottish management, had one lighthouse for 12.3 miles of coast. The Royal Commission also appreciated that the extensive, convoluted mileage of the Scottish

Corran Narrows

Bailey Head, Dublin, Eire

Rubha nan Gall lighthouse, Mull

coast did not require so many lighthouses, they still concluded it was insufficiently lit. The Commissioners report was finally published in 1861 – and totally failed to ignite any serious reforms. Happily the NLB was to retain some sort of autonomy, having opposed any form of centralisation from the outset.

During this turbulent episode the NLB had managed to exhibit the first five lights sanctioned by the Board of Trade – Rubha nan Gall, Ornsay, Kyleakin, South Rona and Ushenish on South Uist – all on 10 November 1857. Two other major lights were to incorporate Board of Trade economies. For example, the Butt of Lewis (1862) was built of brick by John Barr & Co of Ardrossan, but it was late in the year 1859 before the materials reached the Butt. The vessel was wrecked whilst unloading cargo so masonry work had to be postponed until the following spring. Thus when the Commissioners visited on 23 July 1860 they found 'work not so far advanced as had been anticipated.' By 24 March 1862 they were told that things had now progressed so well that the lighthouse would be functioning by the autumn. In the meantime the mason employed to build the 168 steps up the 37 m (120 ft) tower had gone on strike until he got a penny a day extra. Furthermore Trinity House and the Board of Trade insisted on a fixed light, rejecting the Stevensons' plan to employ a flashing white light in order to distinguish it from Cape Wrath and Stoer Head. It would be another 43 years before the flashing light came into operation at the Butt. All its supplies were brought by boat to a jetty nearby until 1960 when road access was improved.

Another lighthouse to be built of brick was that of the Monach Isles, in 1864. In Gaelic it is called Heisgeir and to distinguish it from Hyskeir off Canna we will call the former

Left: Butt of Lewis lighthouse, Outer Hebrides

Below: Rounding the Butt of Lewis

Heisgeir Monach, or even Shillay – the island on which the lighthouse is actually built. There are five main islands lying some 5 miles (8 km) off North Uist, with their highest point only 19 m (62 ft) above sea level with rocky shores and skerries of Lewisian gneiss, beautiful shell sand beaches, dunes and machair (or flat fertile pasture rich in flowers and other wildlife). Although windy, experiencing gales in up to 44 days a year, the climate is mild – winter or summer – showing the ameliorating effects of the surrounding Atlantic.

The Monachs archipelago was declared a National Nature Reserve in 1966. The bird list currently stands at around 150 species, but only 38 or so are breeding birds. Quite a few migrants were recorded by the lightkeepers over the years. Plant and invertebrate lists are also quite respectable for such isolated islands but there are few mammals recorded. Rabbits were introduced as food, and then feral cats to control them – mercifully the latter are now extinct but probably wreaked havoc with ground-nesting birds. Grey seals breed in the autumn and, since abandonment by humans, the islands now support the largest colony in Britain, second only in the world to Sable Island in eastern Canada.

The human population peaked at 135 in 1891 but declined dramatically thereafter. Captain Otter knew of many vessels to have foundered around the islands in the first half of the 19th century and recommended that a lighthouse be built. The NLB Commissioners agreed and in 1862 bought the island of Shillay from North Uist Estate (who still own the rest of the archipelago) for £400. It was here that David and Thomas Stevenson came to build the light at a cost of £14,673: it was first exhibited on 1 February 1864. The 41 m (133 ft) tower was of red brick and reputedly built on exactly the same spot that the monks had maintained a coal-fired beacon in medieval times.

In 1886, the first year of the new light's operation, the apparatus that turned the light broke down, and the three keepers (Principal Lightkeeper Robert Seater, Assistant Lightkeeper Lachlan Campbell and an occasional) kept it turning by hand for no fewer than 17 nights. The master of the Pharos timed their efforts and remarked at their astonishing degree of accuracy!

On 15 November 1936 a calamity occurred at Heisgeir Monach not dissimilar from that at the Flannans. Principal Keeper JW Milne and Assistant W Black took a dinghy from Shillay over to the neighbouring island of Ceann Iar as they did regularly. Their task was to collect supplies and mail from the neighbouring tidal island of Ceann Ear – a walk of nearly three miles across two fords. The weather had deteriorated badly by the time they returned. They were swept to their deaths by strong currents and high winds. Monach, at the time, was manned by only two keepers and it is said that Milne's sister had to report the situation by radio and maintained the beacon herself overnight until help came the following day. The keepers' bodies were washed ashore on the islands three weeks later.

The Monach light was discontinued in 1942 and it was only months afterwards that the last inhabitants abandoned the Monach Isles altogether. Shillay is still owned by the NLB, the other islands in the group by North Uist Estate. When the islands became a National Nature Reserve in 1966, the NLB offered Shillay on a 99-year lease (expiring in 2063) on an annual payment of ten shillings (50 pence in today's money) 'if asked'.

*Grey seals breeding on the
Monach Isles*

*Left: Heisgeir lighthouse, Monach Isles, North Uist
Right: Heisgeir Monach lightroom*

But another challenge had too long remained in the shadows – the vicious Torran Rocks, 18 km (11 miles) south of Mull. Fourteen kilometres (8 miles) beyond, they terminated in a lethal outcrop of Dubh Heartach, or Dubh Artach as it is now known, or as Robert Louis Stevenson had it: 'only the first outpost of a great brotherhood – the Torran Reef that lies behind it.' His father, Thomas, first landed on the rock during his inspection voyage of 1864 and resolved that, since it lay in the track of shipping heading for the Irish Sea, it badly needed a lighthouse.

The urgency of the task was highlighted in 1865 after severe winter storms in the vicinity resulted in 24 vessels foundering in little more than a month. The rock is pounded with waves built up across 3,200 km (2,000 miles) of open Atlantic and channelled to spectacular proportions at Dubh Artach by a 130 km (80 miles) submarine valley which Robert Louis surmised accounted 'for the seemingly abnormal seas to which the tower is subjected … ' The Stevenson brothers realised that a tower of utmost strength was required on this rock only 10 m (30 ft) high. They estimated it would cost £56,900 and were given the authority to proceed in spring 1867.

Now David and Thomas Stevenson were about to tackle the equivalent of their father's Bell Rock, and their elder brother's Skerryvore, just 34 km (21 miles) to the north. Robert Louis described it thus:

> An ugly reef is this of the Dhu Heartach; no pleasant assemblage of shelves, and pools, and creeks, about which a child might play for a whole summer without weariness, like the Bell Rock or the Skerryvore, but one oval nodule of black-trap [basalt] sparsely bedabbled with an inconspicuous fungus, and alive in every crevice with a dingy insect between a slater and a bug. No other life was there but that of seabirds, and of the sea itself, that here ran like a mill-race, and growled about the outer reef for ever, and ever and again, in the calmest weather, roared and spouted on the rock itself.

By September the work force had managed only 27 landings before bad weather drew a halt. A wharf was constructed at the shore station on Earraid, 23 km (14 miles) away, where the Mull granite would be sourced. On the rock itself the iron foundations for the workmen's barrack were set in position. Thirty-eight days' work was achieved the following season but the two-storey iron barrack was complete. On the last day Alan Brebner, the foreman, decided to stay on with 13 men to take advantage of the settled weather; a storm brewed up, waves burst open the door of their refuge, where they had to remain for five days before rescue. David was not amused, though pleased to see them safe. 1869 saw just 60 landings saw the stone foundations laid to a third course, all set in Smeaton's Portland cement, but a storm on 9 July washed away 14 of the two-ton blocks, 11 of them into the sea. The fury of the gales that summer forced the solid stone base to be heightened from 14 to 21 m (45 to 70 ft) (Skerryvore necessitated only 10 m.)

On calm days Robert Louis observed:

Pier, harbour, quarry site and observatory at Earraid, Isle of Mull

This small black rock, almost out of sight of land in the fretful, easily-irritated sea, was a centre of indefatigable energy. The whole small space was occupied by men coming and going between the lighters or the barrack and the slowly-lengthening tower … All the time, the sea would be heard roaring over the external reef, and even from the seaward side of the rock itself, there would come every now and then, a report and a column of white foam. Such was the life on Dhu Heartach in calm weather – a burst of activity, perhaps sixteen or seventeen hours a day … In storms … they sat listening to the thunder of the seas …

The master builder was a veteran of the Stevenson work force, Robert Goodwillie. In storms like these, Robert Louis recalled: 'I see before me still in his rock-habit of undecipherable rags [Goodwillie] would get his fiddle down and strike up human minstrelsy amid the music of the storm.' He kept a detailed log of his time on the rock, running to more

Dubh Artach lighthouse (left by Calum Falconer)

than 400 pages which, through Robert Louis' stepdaughter, Isobel Field, have come to be preserved in California. Every day he recorded the weather, sea state, whether he could see Skerryvore or Ruvaal lights and the mainland hills, and of course the day's building progress.

At Dubh Artach at low tide the landing stage is 10–11 m (35 ft) above a boat, yet not completely out of the reach of the swell. Landings other than via the precarious use of dangling ropes from a derrick were most unusual even on calm days. Robert Louis himself managed it on 8th and then again five days later, while on 19 August 1870, between 12.40 and 1.40 p.m. Goodwillie recorded that Robert Louis' 41-year-old mother landed – 'the first woman we have had on the Rock.' She may well still be the only woman who has been ashore there. Robert Louis specifically noticed the rock 'alive in every crevice with a dingy insect between a slater and a bug.' Keepers would soon learn to keep the lights on in the tower, for the sea slaters, some up to a foot (30 cm) long, would invade at night and attack food stores.

Thirty-one courses of rock had been completed in 1869 and the full height of the tower, 38 m (125 ft) (4 m short of Skerryvore), was completed by the fourth season (1871), involving 71 courses totalling 3,115 t of stone. The fifth season, with Thomas Stevenson as site engineer, saw the assembly of the lantern and the installation of the optical mechanism plus internal fixtures and fittings, so that on 1 November 1872 the Dubh Artach light was finally exhibited. James Ewing, the first principal lightkeeper reported back to base: 'I beg leave to inform you that the light exhibited on the 1st conforms to your orders. I am also glad to state that we carry a magnificent flame, which eventually must eclipse all the lights on the west coast.' It is said that the light was so bright that it affected the eyes of the keepers, and a month later they requested to be supplied with 'preserves' – sunglasses!

Within five days of being lit however, a furious storm had ripped off the copper lightning conductor; but the tower survived the winter. The barrack was then dismantled, leaving only a few iron stanchions. The final accounts came to £65,784, nearly £9,000 over estimate while the completed shore dwellings and wharf added a further £10,300. Dubh Artach became the third pillar and rock lighthouse in Scotland, all three being remarkable achievements in their own right. Yet is should be remembered that the first, Bell Rock, was built on a largely submerged reef, by Robert Stevenson and his work force through hard manual labour, using simple hand tools, employing only rowing boats and sailing ships!

James Ewing looked after the Dubh Artach light for the next 11 years. Despite the exceptionally adverse conditions faced by the keepers, which resulted in them receiving additional payments in kind, Ewing was not the only one who saw long service at this unpopular posting. Some, however, found the lonely rock and its cramped quarters less to their taste; one even had to be prevented from diving into the sea and attempting to swim ashore.

Notwithstanding additional incentive payments, the remote location of rock lighthouses actually suited some veterans. Archibald McEachern was assistant lightkeeper at Skerryvore for 14 years from 1870–84 and John Nicol was its principal from 1890–1903. The latter was involved in a dramatic rescue when the liner *Labrador* en route from Halifax, Nova Scotia, to Liverpool ran aground on the nearby Mackenzie's Rock in 1899. The lifeboats were manned and two made it to Mull, but one with 18 passengers reached the lighthouse where they were looked after for two-and-a-half days before they could be taken to the mainland. No lives were lost and Nicol and his two assistants were commended by the NLB for their efforts.

In order to distinguish Dubh Artach's tower from its neighbour Skerryvore, 34 km (21miles) to the north, a red stripe was added to the former's white, wave-washed livery in 1890. Notorious for overdue reliefs, in January 1947 two light aircraft had to drop emergency rations to stranded keepers. The lantern equipment has since been modernised and then converted to automatic operation in 1971. Not surprisingly, given the difficulty of access and its unpopularity as a posting, Dubh Artach was one of the first rock lights to be automated. For some time the families of both Skerryvore and Dubh Artach keepers were domiciled at Earraid until finally transferring to Oban.

In 1872 David A Stevenson, Thomas's nephew, was to deem the sturdy, functional tower of Dubh Artach as: 'One of the best architecturally in the service suiting as it does so

NLB shore station on Earraid, Isle of Mull

Chicken Rock and the Calf of Man lighthouses, Isle of Man

well the rock on which it stands.' Notwithstanding, David and Thomas had already started another, but submerged, rock tower, further south, off the Isle of Man – Chicken Rock. Two lighthouses already stood on the dangerous Calf of Man, a tiny island off the southern tip of the Isle of Man itself. They were arranged in line, one above the other, to mark the position of Chicken Rock but both could be obscured in the sea mists that were common in the area.

Chicken Rock lay only three-quarters of a mile (1.2 km) offshore, an extensive reef that showed only a metre (3 ft) above the surface at high tide, creating a highly dangerous obstacle. It had gained its name from a tiny seabird, the storm petrel, known to superstitious sailors as 'Mother Carey's Chicken'. Obviating the need for a barrack on the rock, a convenient shore base already existed close by at Port St Mary where the lighthouse granite would be delivered from Dumfriesshire and finally dressed. In 1869 work progressed so fast out on the rock that the foundation pit was excavated in the first season, nine courses of stone were laid in the second and the rest of the solid base (23 courses to a height of 10 m or 30 ft) completed in the third. The final 96 courses were finished by June 1873, after which the craftsmen and joiners moved in to complete the interior and the lamp was assembled. At first Trinity House had insisted, against the Stevensons' better judgement, upon the design of the light system and the incorporation of a red arc, which would prove of little value in hazy conditions. They went ahead however, under protest, and were soon to be proved correct, forcing Trinity House into an embarrassing climb-down so that the Stevensons' original design finally could be installed. In a flurry of snow squalls the lamp was lit on 1 January 1875 and both beams on the Calf of Man (erected by Robert Stevenson in 1818) were extinguished for good.

Chicken Rock and the Calf of Man

Chicken Rock performed well, until, on 23 December 1960 an explosion occurred in one of the storerooms. Unable to reach the radio room below, the three keepers had to retreat up to the lantern from where they then lowered themselves by rope to the rock outside. Fortunately the smoke was spotted from the shore and within half an hour the

Port St Mary lifeboat arrived. Rocks and strong tides made it difficult for them to get a rope ashore and when they did manage to rescue the keeper in most need of medical attention he was dumped in the sea, just to add to his troubles. When he was safely aboard, the lifeboat crew deemed it urgent to get him to hospital immediately. They had to leave the other two keepers on the rock for eight hours in all, on a rising tide. A helicopter overhead was helpless but remained on station with another lifeboat until the first lifeboat returned. With great heroism, and a slight improvement in the sea conditions, they succeeded in taking off both men. A truly remarkable rescue.

Shortly afterwards the NLB tender *Hesperus* put a flashing buoy in place beside the now unmarked reef, and the opportunity was taken not just to repair the Chicken Rock tower but also to convert it to automatic operation. After the fire at Skerryvore in 1954 an automatic Dalen light was placed on the rock until the tower was repaired; the success of this Swedish invention (a valve triggered by the sun to turn on bottled acetylene gas) pointed the way for future automation, for example when Chicken Rock came back into service in September 1962: solar panels were installed in 1999.

Taiaroa Head lighthouse, near Dunedin, New Zealand

By now David and Thomas were advising on projects in Japan, India, Newfoundland and, later New Zealand. Thomas's wife Margaret had a brother James Melville Balfour (1831–69) who served his time as a civil engineer in the Stevensons' firm. In 1863 he was sent to New

*Akaroa lighthouse
relocated, New Zealand*

Zealand to advise the Otago Provincial Government and brought lamp equipment he had designed for Cape Saunders and Taiaroa Head lighthouses. I visited Taiaroa in February 2013 to view the royal albatross colony but did not have access to the lighthouse. A few days earlier however I had visited Akaroa further north. The old wooden lighthouse had been dismantled from a nearby headland in 1980 and reconstructed on the outskirts of the town. Shown round by volunteers I was delighted to see the brass plaque on the optics showing it had been 'Designed by D and T Stevenson and James Dove & Co Manufacturers Edinburgh 1878'. The light was exhibited on 1 January 1880.

Akaroa lighthouse plaque, New Zealand

Balfour, uncle to Robert Louis Stevenson, had remained in New Zealand to marry, and was appointed colonial marine engineer with a special responsibility for lighthouses. Sadly on 18 December 1869 he was drowned while on his way to attend the funeral of a friend. Fellow engineers remembered him as 'a far-seeing man of boundless energy and sound judgement whom the young colony could ill afford to lose.' As I saw for myself in Akaroa, and we will discover later at the Stephens Island lighthouse, the Stevensons continued to advise on lighthouse development in New Zealand and elsewhere.

The Stevensons were also busily engaged in building harbours and breakwaters in Berwick, Eyemouth, Peterhead and Wick. For three years a reluctant Robert Louis Stevenson served under his father as an apprentice engineer. He had landed on Muckle Flugga in 1869 and spent three weeks on Earraid in 1870. Two years before, he had worked for a month on the Wick breakwater where he had donned a diver's suit for the first time. Later Robert would comment:

Diving was one of the best things I got from my education as an engineer: of which however, as a way of life, I wish to speak with sympathy. It takes a man into the open air; it keeps him hanging about harbour sides, which is the richest form of idling; it carries him to wide islands; it gives him a taste of the genial dangers of the sea; it supplies him with dexterities to exercise; it makes demands upon his ingenuity; it will go far to cure him of any taste (if ever he had one) for the miserable life of the cities. And when it has done so, it carries him back and shuts him in an office!

Much to Thomas Stevenson's dismay the Wick breakwater was destroyed in a violent North Sea storm a few months later – according to Robert 'the chief disaster of my father's life.'

Although he had presented his first professional paper *Notice of a New Form of Intermittent Light for Lighthouses* to the Royal Scottish Society of Arts in March 1871, Robert Louis Stevenson announced only two weeks later that he was abandoning engineering. Surprisingly his mother described Thomas as being 'wonderfully resigned' to the decision. In truth, despite his own boyhood dream of being a writer, Thomas was sorely upset at his only son's decision. Several months later young Robert's paper won one of the Society's silver medals so he was able to say to his parents: 'No one can say that I give up engineering because I can't succeed in it, as I leave the profession with flying colours.' He then went on to prepare, in conjunction with his father, a second paper for the Society on *The New Lighthouse on the Dhu Heartach Rock, Argyllshire* but it was never completed; the 4,800-word manuscript (lodged in California) was finally published, as only 1,000 copies, in 1995.

Although David and Thomas's firm was engaged in many diverse engineering projects, lighthouse work was not neglected of course. The annual inspection cruise continued, but now involved ever more sites, the Stevensons ignoring no detail however small. They were also sensitive to the needs of their employees and so were always highly regarded by lightkeepers. As a result the keepers were always keen to participate in data collection, such as meteorological recording, observations on waves and natural history. On 6 January 1874 the lightkeepers on Dubh Artach reported an earthquake shaking the rock and the tower. The keeper at Lismore, on the Great Glen Fault, experienced another on 23 April 1877. No damage was done in either case.

At this time experiments were begun with electricity as a power source, encouraged by the Board of Trade who recommended the first trials were undertaken on the Isle of May, an unusual choice, adding expense due to the isolation. They may have been influenced by the recent spate of shipwrecks in the vicinity of Fife Ness which prompted the mooring of the North Carr lightship. This obviated the need for the Low Light on the Isle of May built by Alan Stevenson in 1844, and now the island's Bird Observatory.

The main, highly ornate building erected on the Isle of May by Robert Stevenson in 1816 incorporated rooms for visiting officials, but could only accommodate three keepers and their families. The new apparatus now required six keepers, together with engine house, boiler house, chimney, workshop and coal store. These were built in a small valley containing a small freshwater loch. The carbon-arc lamp produced a brilliant three million candlepower; with a refinement by Thomas Stevenson permitting it to be tilted downwards in poor visibility, better to illuminate the water surrounding these dangerous shores. At the end of the previous century Thomas Smith's pioneering beams were only 30 candlepower. The new May light was visible for 22 miles (35 km), 40 or 50 miles (64–80 km) the flashes lighting up the clouds overhead! A paraffin oil lamp stood by as a backup. The whole operation cost £22,000 with maintenance three times the cost of a paraffin light. As a result, and with safety fears and the increased efficiency of incandescent mantles on oil lamps, the Isle of May apparatus was discontinued in 1924. In 1972 the families moved across to live at Granton and, with automation, the last keepers left in 1989 – ending a run of no fewer than 353 years of diligent attendance.

All the time both David and Thomas led full social lives at home in Edinburgh – spending time with their families, giving lectures, writing scientific papers and reports, supporting local societies and their church. A highly skilled and experienced engineer, David seems to have found time for fewer, less diverse interests than his father and elder brother. Nonetheless, being a natural writer he took copious notes on his travels, beginning with a diary his father required him to keep on his first lighthouse voyage, aged 13. On 18 August 1828 the 141 ton *Regent* departed Newhaven to land first on Inchkeith. The party were shown around by the principal lightkeeper, a Mr Bonnyman, who had lost a finger as a mason when the Bell Rock was being built. They also met an old lighthouse pilot called Noble, who had had 20 children. The next day they landed on the Isle of May, where their pointer dog was killed whilst scrambling for birds' nests on the steep cliffs.

> We contrived … by shouting, clapping hands, and throwing stones, to put up thousands of seagulls to flight [Hardly acceptable behaviour nowadays!] But, although they almost darkened the sky, Mr Pithie the lightkeeper assured us that the major part of the birds had left the island with their young, so that what we saw were chiefly the herring and laughing [?] gulls, the Marrot [guillemot] and Picktarnie [tern] having already migrated.

Young Thomas's performance at school had been unremarkable, although he enjoyed Latin and English literature. Indeed he began working life in a printing firm owned by a friend of his father and produced some of his own essays on his own little printing press. But in the end, aged 17 he had entered the family business, continuing to write mostly for engineering and architectural journals, advocating the blasting of useless ruined buildings and restoration of historic ones. In 1845 he delivered to the Royal Society of Edinburgh a paper on forces exerted by sea-waves, the first significant work on this subject. Then followed some 60 published articles on such diverse topics as defects in rain gauges, the geology of Little Ross Island and of course engineering topics. All demonstrate, not just his communicative skills, but his facility for observation, experiment and quantitative analysis of natural phenomena and man-made constructions. He was a founder member of the Scottish Meteorological Society in 1855, was elected President of the Royal Society of Edinburgh in 1864, and became a Fellow of the Geological Society 10 years later.

In 27 years David and Thomas's firm amassed fees amounting to £90,000 (after salaries and expenses), one quarter of which came from business for the NLB; this gives some indication of the diversity of projects with which the brothers' firm was involved. Between them they completed some 28 major lighthouses between 1854 and 1880. Not surprisingly, with this arduous workload both David and Thomas had health problems. In 1881 David became so ill that he could no longer work and finally retired in 1883, leaving Tom to take over – until he too began to suffer.

After a long and illustrious career, a most effective manager for the family firm, David died in 1886. He was a man of sound judgement, upright, kind, open and approachable with wide interests, in railways, sewerage system, agricultural implements and collecting artworks, he had also travelled widely abroad. He had four daughters and four sons, but only two of the latter (David Alan and Charles) survived – indeed both to become lighthouse engineers.

The year David that died, the NLB were operating no fewer than 68 manned lighthouses. Alan Brebner had been made a full partner in the business. David's two sons also began to assume effective roles, though Thomas always retained for himself the controlling shares. Indeed Thomas was reluctant to reward Charles, the younger, any respectable share. He lamented that he would not be handing over to his own son, and it would remain a perennial source of bitterness for him that Robert Louis had proved totally disinterested in engineering.

Then Thomas was mortified to find his nephew David Alan (David A.) had been appointed as joint NLB engineer alongside him. Together they would be responsible for

Fidra lighthouse, Firth of Forth

only three lights. Fidra in the Firth of Forth, was reluctantly sold by Dirleton Estate to the NLB in 1885 to allow the lighthouse to be built. The Estate was still nursing a grudge when it tried to sue for the intrusion of the foghorn; the judge ruled that a noisy foghorn was more agreeable than a quiet shipwreck. In 1970 Fidra – again, a lighthouse conveniently close to Edinburgh – became the first major offshore light to become automatic.

The very first NLB installation to be automated however (as early as 1894) had been Oxcars, a two-man light built in 1886 in the Firth of Forth. Thus began a programme of rationalisation: Ornsay and Kyleakin were reduced to one-man operations in 1898, Chanonry, Cromarty and Rubha nan Gall by 1900. David A and Charles then began to consider the possibility of unmanned gas-lit boats to exhibit a stronger light at a greater height above the sea than was possible on a buoy. Sanctioned by Trinity House and the Board of Trade, the first of its kind in British waters was the Otter Rock light vessel moored off southeast Islay in 1901. Others were to prove less successful, yet four remained on station for nearly 50 years.

Ailsa Craig or 'Paddy's Milestone' as it is nicknamed, was purpose built in 1886 for gas operation, with the gas being produced on the island by heating mineral oil with coal. Gas-operated lights did not prove popular with the NLB; although bottled gas was used in some instances, the Stevensons feared the flame could be extinguished by sea water, and any connection to domestic supply near a town could prove unreliable. Incidentally, Ailsa was also the place where communications between the keepers and the shore were tried with homing pigeons; but they were easily picked off by the local peregrine falcons!

Relationships between David A and his uncle became very strained. Thomas became somewhat irrational and the situation was to take a further toll on his health until he died in 1887, only a year after his brother. Perhaps, in atonement for failing to follow in his father's footsteps, in 1886 Robert Louis Stevenson was moved to write of him:

> … he, two of my uncles, my grandfather, and my great grandfather in succession have been engineers to the Scotch Lighthouse Service; all the sea lights of Scotland are signed with our name, and my father's services to lighthouse optics have been distinguished indeed. I might write books till 1900 and not serve humanity so well; and it moves me to a certain impatience, to see the little, frothy bubble that attends the author his son, and compare it with the obscurity in which that better man finds his reward.

For all their ground-breaking contributions over the years to, for example, lighthouse design, oil lamps and optics, and foghorns, the Stevensons – and other workers in the field – never patented any of their inventions. They could have made themselves fortunes but saw their work as for the greater good of humanity. Robert Louis reflected: 'The Bell Rock stands monument for my grandfather, the Skerry Vhor for my Uncle Alan; and when the lights come out at sundown along the shores of Scotland, I am proud to think they burn more brightly for the genius of my father.'

Ailsa Craig and its lighthouse from the air (Tony Mainwood)

The final word, from a book dedication, again by his son: 'Thomas Stevenson, Civil Engineer, by whose devices the great sea lights in every quarter of the world now shine more brightly.'

nine

The Works of my Ancestors

Whenever I smell salt water, I know that I am
not far from one of the works of my ancestors.

Robert Louis Stevenson, 1880

David Stevenson's two surviving sons, David Alan or David A (1854–1938) and Charles Alexander (1855–1950) inherited the business. They were different types, just as their father and uncle had been. Both had been groomed by their father, well educated with Bachelor of Science degrees from Edinburgh, widely travelled and enjoying many and varied interests, both professionally and in their private lives. From the age of 17, under the guidance of his father and of Alan Brebner (1826–90), David A had developed an absorbing passion for engineering. At 20 he had spent a summer at Earraid while Dubh Artach was being built and gained a commendation from the Institute of Civil Engineers for delivering a paper on the progress; another paper on the Ailsa Craig lighthouse won him their Telford Prize. He also addressed the Institution of Mechanical Engineers on the installation of the electric light at the Isle of May. With his brother, he revised their father's influential textbook *Canal and River Engineering* and acted as extramural examiner for Edinburgh University. David also gave evidence before royal commissions, parliamentary committees and in important legal cases. He followed his father's business interests in insurance and, with his brother, was elected a Fellow of the Royal Society of Edinburgh along with numerous other engineering institutions. Both boys shared their father's interest in archery, while Charles also enjoyed skating, horse riding, shooting, cycling and golf.

Charles was only a year younger but was much less organised, indeed impetuous. He corresponded with *Nature* and other scientific journals on such diverse matters as the 1889 earthquake in Scotland, seismography, river discharge formulae, wind velocity and dioptric lenses, while also publishing on many technical aspects of engineering. He had an alert and inventive mind, with an intuitive approach to engineering. In 1877, in America with his brother, he had seen a compressed air foghorn in operation and immediately saw its application for the NLB. A short time later he managed to have two sent to him in Scotland

– the ones that were to be installed at St Abb's and on Sanda. He had seen Alexander Graham Bell's telephone and began to ponder how it could be altered to operate without wires. The problem of how it could provide communication from lightkeepers on remote stations such as Muckle Flugga and Skerryvore obsessed him and it is said that he came up with a system a year or two in advance of Marconi. The problem of poor and unreliable navigation buoys also intrigued him, eventually feeding his uncle a simple yet highly effective solution. Thomas had always been somewhat reluctant to accept his younger nephew and Charles only started receiving a salary after his father died in 1883. In 1887 when Thomas died David A, Charles and Alan Brebner were able to carry on, without interference, as partners in the family firm.

David A immediately stepped into his uncle's shoes as NLB Engineer, a post he was to hang on to tenaciously until his final retirement in 1938 and ending 150 years of family involvement. David A only had daughters and maybe, with his reluctant attitude to Charles's son David Allan (D Alan), there was a similar element of resentment that Thomas had shown about handing over to his nephew, David A himself. Charles's son D Alan (1891–1971) was quick to reveal an aptitude and passion for engineering, lighthouses in particular, so was the obvious heir apparent to the firm. With no official role within the NLB, Charles applied himself to civil work, and was happy to iron out problems for his brother who had to concentrate upon lighthouse work with its associated paperwork and report writing. Charles, notwithstanding his uncle Thomas, was undoubtedly the most inventive member of the family so, together – just like their father and uncle – brothers David A and Charles complemented each other into a very effective team.

Financial restraints in the 1880s had made the Board of Trade reluctant to take on new lights so none showed in Scotland between 1886 and 1892. Yet they readily sanctioned the costly electrification of the Isle of May and the placing of the North Carr lightship. In 1894 a General Lighthouse Fund was finally adopted whereby all light dues were directed towards the maintenance of lighthouses – although, frustratingly, still requiring the go-ahead from Trinity House. Congestion and collisions in the English Channel were encouraging more

Fair Isle from the south

and more ships to take a northabout route to and from the North Sea. Ships favoured the Sumburgh passage rather than face the Pentland Firth. Between Orkney and Muckle Flugga there was only Sumburgh Head to mark untold unmarked dangers. It was obvious that more and better navigation lights were required. Thus the first crop of lighthouses the brothers tackled was in the north of Scotland, some of them as challenging as anything gone before.

Fair Isle lay midway between Orkney and Shetland and a lighthouse was suggested as long ago as 1809, with a firmer proposal being refused in 1867. In 1870 Robert Louis Stevenson commented: 'The coast of Fair Isle is the wildest and most unpitying that we have seen. Continuous cliffs from one to four hundred feet high tower by huge voes and echoing caverns … in a belt of iron precipices.'

Fair Isle North Light and cliffs

David and Thomas had visited Fair Isle in 1877, recommending a light and fog-signal at each end but finances precluded any further action. An experimental fog warning system using rockets, was installed in 1885, but operated for only a year. It would be 1890 before the two lighthouses were begun, with David A cutting the first turf for the South Light. The workforce, mainly from the mainland, found the situation particularly lonely so islanders had to be employed as well (only to disappear as soon as the fishing season resumed!). Masons arrived from Aberdeen in March 1891 and the 26 m (85 ft) tower was completed by 5 June, and first lit on 7 January 1892.

The North lighthouse, 14 m (46 ft) high on a 75m (246 ft) cliff, came into service on 1 November that same year. Here the Stevensons had installed their first innovative 'hyper radiant' light, devised by Thomas and improved by Charles. Larger burners were used to increase the strength of the beam. Such was the intensity of the light source that the lenses had to be placed further away to avoid overheating. The North Light became a rock

Fair Isle
North Light

1891

station in 1978, when the families moved to Stromness, and it was eventually automated in 1981 with an electrically driven revolving system and battery-operated lamps. A computer known as 'Fiona' alerted the South Light keepers of any malfunction, by then a rock station also.

One keeper complained that Fair Isle was the most remote shore station he knew: the mainland being a long way away to spend his leave. 'But now it's a rock station it's probably about the best in the service' he added. With only 60 or so inhabitants the six lighthouse keepers and their families contributed much to the little island community, especially numbers at the local school. In 1998 Fair Isle South became the last Scottish lighthouse to be automated, following a long programme of conversion begun in 1960 (see later).

Following the grounding of a steamship in 1891, a lighthouse was completed on Helliar Holm, a small tidal island near Shapinsay in Orkney in 1893. The island is uninhabited by people but is grazed by a flock of seaweed-eating sheep introduced from North Ronaldsay. A Panamanian container ship dragged her anchor and grounded on Helliar Holm in February 1999.

Lying 55 km (34 miles) west of Orkney and the same distance from the Sutherland coast, Sule Skerry was to become the most remote light station in Britain. Its 14 ha rise to no more than 14 m (46 ft) above sea level and landing is difficult, so this was no place for families to live. Substantial houses were built for them in Stromness, where the NLB and its tender, until recently, had a base. Keepers could be stranded for up to a month in bad weather so it is not surprisingly that this was one of the first lighthouses to have radio communication. Bad weather and short days precluded any construction work in the winter so the 27 m (89 ft) tower and associated buildings took two seasons to build. This was done by John Aitken of Lerwick who had no experience of lighthouse building but kept his estimate low by convincing the skipper of the weekly steamer between Stornoway and Stromness to drop off materials as it passed. But the most surprising innovation of all – and long overdue – is best described by David A Stevenson himself:

> I have introduced for the first time also, what I think will greatly conduce to the efficiency of the light, and to the comfort of the keepers, namely a handrail round the balcony which will enable the keepers even in the worst of weather to go outside and remove snow and ice from the glass without fear of accident, and it will further be a great advantage for the daily cleaning of the lantern panes.

It is hard to believe that keepers had to fulfil these tasks all this time without any handrail! Final illumination was delayed until 1895 while Trinity House and the Board of Trade argued with the NLB about the cost and the character of the light. A hyper-radiant system was used, the same as the Fair Isle North Light. Sule Skerry was automated in 1982 but it is still occupied by thousands of puffins, grey seals (the great selkies of Sule Skerry, according to the folk song) and a small number of breeding gannets from its parent colony on the tiny Sule Stack or Stack Skerry about 6 km (3 miles) to the southwest.

Sule Skerry, west of Orkney

Sule Skerry lighthouse, meteorological station and chopper

The Stevensons' next challenge was an unusual one – building a lighthouse at Rattray Head lying between Fraserburgh and Peterhead. For decades Trinity House and the Board of Trade had blocked such a venture, deeming that Kinnaird Head and Buchan Ness were sufficient navigational aids round this stormy corner of northeast Scotland. In a final act of desperation David A collected a petition with the signatures of, amongst others, local fishermen, shipping companies and their skippers, the Navy and local pilots, before the London powers gave way. By April 1892 a quarry was found at Rubislaw in Aberdeen and work got underway – although a terrible storm washed away several blocks from the first three courses. The design was novel, stipulated by the tower's unusual situation half a mile (less than a kilometre) offshore. The gap was tidal so the reliefs could be achieved by tractor at low spring tides, with a final climb up a vertical 10 m (33 ft) ladder to the entrance door. The problem was that there was nowhere convenient to put the foghorn, so Charles Stevenson built a substantial concrete base to accommodate the machinery – the first time a siren fog-signal had been installed in a rock lighthouse. The foghorn itself sat on top of the substantial basal structure known as 'the quarter deck', alongside the 34 m (112 ft) tower itself which in 1895, after three season's work, flashed 44,000 candlepower, compared with only 6,500 at neighbouring Buchan Ness.

Rattray Head lighthouse between Fraserburgh and Peterhead

At this time too, the Stevensons were tackling a new light in the Pentland Firth, on the island of Stroma two miles (3 km) off the Caithness coast, inhabited by over 300 people. A minor light had been placed here in 1890 but the manned light came to be built in 1896 on Swelkie Point, immediately above the notorious tide rip that gave it its name. Stroma is a Norse word meaning 'the island in the stream' and in 1814 the Swelkie led Sir Walter Scott to call Stroma 'a wicked little island'. In April 1910 a young keeper fell to his death while washing the glass panes of the 23 m (75 ft) tower. On 22 February 1941 a German plane sprayed the lighthouse buildings with machine gun fire, narrowly missing the children playing outside.

The damage was easily repaired. In 1961 it was decided to make Stroma a 'rock station'. By now there were only a dozen people living on the island so the loss of the lighthouse families to Stromness may well have been a contributory factor in the ultimate abandonment of the island the following year. In 1972 helicopter reliefs were introduced and the families moved from Orkney to Scrabster. That year also the light was made electric with a sealed beam unit. Stroma light was eventually automated in 1996.

Above: Stroma lighthouse, off Caithness, in the Pentland Firth

Below: Stroma's old foghorns

Stroma lightroom backlit

The third major challenge for the Stevensons however, was the construction of the lighthouse on the Flannan Isles. But by now David A was gravely ill with a nervous complaint. He had long periods of convalescence during which Charles covered for him with the NLB; Charles was also pursuing a bewildering host of other inventions and engineering projects both at home and abroad. Nonetheless, liaising closely with his invalid brother, Charles took on this new task.

The Flannans lie some 32 km (19 miles) west of the Isle of Lewis in the Outer Hebrides. With increasing transatlantic shipping through the 19th century, in 1853 and again in 1880 submissions were put to Trinity House for a lighthouse to be built on the Flannans, but both had been rejected. David A resubmitted the application in 1892 which was also thrown out until he put his case to the Board of Trade and, remarkably, permission was at last granted.

The landing stage on Eilean Mor, Flannan Isles

The largest island Eilean Mor is only 11 ha so it was decided that the lighthouse would be a rock station with a shore base for the families at Breasclete in Lewis. David A drew up his plans and a site was chosen on the grassy summit of the island 90 m (295 ft) above sea level. But the operation would not be easy; the island is surrounded by near perpendicular 50 m (164 ft) cliffs so everything was brought in by sea and had to be hauled up to the site. Two landing stages were prepared, east and west, from which giant flights of steps had to be hewn out of the rock. A horse would be slung ashore by crane to haul everything from each landing place along a tramway at the top.

Sanction had been received from the Board of Trade in 1896 for erection of the Flannan Isles lighthouse and on 7 December 1899 the lighthouse, designed by D Alan Stevenson, was completed by George Lawson of Rutherglen at a cost of £6,914:1:9, which included the building of the landing places and stairs on Eilean Mor. Lawson also built the dwelling houses for the lightkeepers' wives and families at the shore station at Breasaclete, Isle of Lewis, which cost a further £3,526:16:0. The site for the shore station at Breasaclete was chosen for its close proximity to Loch Roag, a sea loch, which provided a safe anchorage and shelter for the lighthouse tender when taking on or putting ashore the lightkeepers, or when bad weather made it impossible to carry out the relief on the due date.

As there was no radio communication between the Flannans and Lewis at that time, a gamekeeper, Roderick MacKenzie, was appointed observer to the light, for which he received payment of £8 per annum. MacKenzie's duties involved watching for any signals from the lighthouse 29 km (18 miles) northwest of his vantage point at Gallan Head, Lewis, and to observe and report any failure in the exhibition of the light. In the event of such a failure he had to report immediately by telegram to Head Office in Edinburgh so that the necessary steps could be taken to have someone sent to carry out any repairs as soon as possible. The station became automatic in September 1971 and the keepers withdrawn. The light was converted from acetylene gas to solar power in 1999 and, like all other NLB lights, is monitored from Headquarters in 84 George Street, Edinburgh.

At the start of the 20th century the Commissioners were keen to fill some of the 'dark blanks' remaining on the Scottish coast. One of the first was Tiumpan Head, Lewis, overlooking the Minch. This had initially been refused by the Board of Trade, although was revisited in May 1879 on being recommended by the Western Highlands and Islands Commission, and from which a watch on illegal trawlers was first authorised. The estimated cost of the lighthouse and building was £9,000 and John Aitken was the contractor. William Frew was appointed as inspector of works. Chance Brothers made the optics and the revolving machine was made by Dove and Co. The light was finally exhibited on 1 December 1900. The lamp was operated from the mains electricity and was a 250 watt mercury vapour type. In case of emergency, such as power cuts, which occur frequently during the winter, there was a 'scheme R' battery-operated light, and in the event of prolonged power failures the paraffin vapour lamp equipment was retained in readiness. The apparatus for driving the lens was a hand wound, weight operated, clockwork machine which had to be wound up every 35 minutes.

The fog-signal was operated by compressed air supplied from a compressor, which was driven by Kelvin Diesel Engine. There were three Kelvin engines and compressors, and when the fog signal was in operation, two of them were in service to maintain the required air pressure with one standby, in rotation. Her Majesty the Queen, with Prince Charles and Princess Anne, visited the lighthouse in 1956 when the seven-year-old heir to the throne sounded the first blast on a new fog siren. There was a complement of six lightkeepers attached to the station: three lightkeepers and their families at the station and a local assistant and two occasional lightkeepers coming in from the village nearby. The Tiumpan lighthouse was finally automated 1985.

Above: Tiumpan Head, Isle of Lewis, at sunset

Below: Tiumpan Head lighthouse

There had been several fatal wrecks in the Firth of Forth in recent years despite the lighthouse on the small island of Fidra having been lit in 1885. So Barns Ness on the mainland further east and the Bass Rock just off North Berwick were sanctioned. The best site on the Bass was on the gun platform of the ruined castle but part of the old fortifications collapsed during the construction so a retaining wall had to be built. The Bass Rock light was exhibited in 1903 and is on the edge of what is now one of the largest gannet colonies in the world. The Bass keepers' families were domiciled in Granton until the light was made fully automatic in 1988.

The following year, 1904, Hyskeir (in proper Gaelic 'Heisgeir' or on some maps – incorrectly – 'Oigh Sgeir') was completed on a low-lying 4 ha skerry of columnar basalt 10 km (6 miles) west of Canna. With an estimate of £15,134, D & J MacDougall of Oban were awarded the contract. The tower was brick-built and 39 m (128 feet) high, its optics being supplied by Chance Brothers of Birmingham (who had long supplied most other NLB lenses) at a cost of £821, the lantern and parapet by Dove and Co., Edinburgh, for another £1,251.

Hyskeir (Oigh Sgeir), off Canna, from the air

Left: Hyskeir (Oigh Sgeir) basalt columns

Right: Hyskeir (Oigh Sgeir) lighthouse garden in 1977

Above: Bass Rock lighthouse from the air. (Stuart Murray).

Below: Bass Rock lighthouse in the old fort with nesting gannets

Bass Rock lighthouse
with its gannets

With their families living in Oban, the duty keepers on Hyskeir came to keep a walled garden on this sea-washed rock, affectionately known as the 'Rock Garden' which grew, amongst other things, wonderful beetroot, cabbages and onions. Other lightkeepers said that it was the most productive garden they had ever seen, and once it even featured on a TV cooking programme. They also kept three goats for fresh milk but by 1977 when I visited the rock, only one called Maisie was left. The keepers were still in post and had created a three-hole golf course! A helipad was laid on Hyskeir in 1973, and its light finally automated in 1997.

In 1938 Hyskeir rock had been bought by John Lorne Campbell as part of his purchase of Canna. In 1907 the NLB had added a small automatic light at the east end of Sanday – an island connected to Canna at low tide, and by a small bridge. Overlooking Rum, Sanday light was modernised in the 1990s.

David A Stevenson now had some 90 lighthouses under his charge (compared with 30 when Skerryvore came on stream) so the NLB inspection voyages visited them on a five year rotation, instead of the previous three. Many lights were undergoing refurbishment at this time, incorporating Charles's more modern lamps. In addition, Maughold Head on the Isle of Man was in progress. In 1904 Charles's son, D Alan, aged 13, was taken on his first inspection voyage and was spending more time in the office learning the ropes. He was a strong boy and heir apparent to the family firm.

D Alan was the seventh of his family to be elected a member of the Institution of Civil Engineers, and the seventh Fellow of the Royal Society of Edinburgh – a remarkable record. He also became a Fellow of the Royal Scottish Geographical Society to serve as its honorary treasurer. Like all his family D Alan developed diverse extra-mural interests: he played the piano and violin and had a special interest in opera; he was also a member of no fewer than six golf clubs and three skating clubs. All his life he maintained collections of pen-nibs, letter seals and stamps (to specialise, after a later visit to South Africa, in Cape triangulars).

The NLB was finding it difficult to experiment with new ideas at this time without seeking sanction from Trinity House and the Board of Trade, so the Stevensons circumvented these issues through the independent Clyde Lighthouse Trust. With strong business representation

Holy Island, off Arran, with its two lighthouses

on the Board, the Trust was always on the lookout for commercial possibilities. As a result Charles installed acetylene gas power at Cloch lighthouse in 1900 with some success; Cumbrae was also fitted with the same system in 1908. With his son, D Alan, Charles also experimented with radiowaves to operate gas-lit floating buoys, and developed the 'Leader Cable' for guiding ships through narrow channels. An electric cable laid on the seabed was detected on board the ship to keep it on track.

The only other lighthouses coming into service before the First World War were Noup Head on the Orkney island of Westray (which in 1908 became the first NLB light to use a system of mercury flotation for the revolving carriage), Holy Island (1905), Neist Point in Skye (1909) and Rua Reidh on the coast of Wester Ross (1912). Neist Point lighthouse looked across the Minch to Ushenish in South Uist which David Stevenson had built half a century earlier. Ushenish was automated in 1970 to be monitored by the keepers in Neist Point 30 km (18 miles) away, until the latter also went automatic in 1990.

Neist Point lighthouse, Isle of Skye, looking towards Ushenish lighthouse on South Uist

When the First World War started D Alan very much wanted to fight in France but his role with the family engineering firm was deemed of national importance. Finally in 1915, to his delight and as a captain in the Royal Marines, he was despatched to Gallipoli to supervise the installation of navigation beacons, erecting five lights in the Dardanelles. He travelled home via Italy and France. Thus by the time the war ended he had gained so much experience that, aged 28, he was made a partner in the family firm – the fifth generation and the eighth consecutive engineer. He did not know then that he would be the last of a long and distinguished line. He married in 1923 and the next year his daughter was born. In 1926 he was approached by Trinity House to become lighthouse engineer in India, where he spent six months. On his return, and now a full member of the Institute of Civil Engineers, he found himself deputising more and more for David A as his uncle became increasingly ill.

*Neist Point lighthouse,
Isle of Skye at sunset*

By now his own father was over 70 yet in 1929 Charles and D Alan collaborated on their greatest invention to date – the 'talking beacon'. A radio signal from a ship received back cross-bearings from two shore stations which then allowed it to calculate its position. The Stevensons improved on earlier American and RAF prototypes, by compensating for the time lag and any distortion from surrounding hills. Despite his stammer D Alan introduced the invention at an international conference in 1929 when, at the same time, the first one was installed at Cumbrae. It became an overnight sensation and was quickly adopted worldwide. Sadly however Britain was slow to take up the idea and Cumbrae remained the only talking beacon until overtaken by radar some ten years later.

Left: Duncansby lighthouse, Caithness

Right: Duncansby Stacks and cliffs, Caithness

After the war came the troubles in Ireland, where three Northern Irish representatives were allowed to sit on a reconstituted Irish Lighthouse Board. The NLB resumed work with lighthouses being built at Duncansby (1924) and Eshaness in Shetland (1929). A temporary light had been erected at Eshaness in 1915 to warn of Ve Skerries, 14 km (8 miles) to the west. Its 12 m (39 ft) tower was square, on top of a 70 m (230 ft) cliff. The station was manned by only a single keeper until it became automated in 1974.

Various other lights were improved, including the Isle of May. Tor Ness in Orkney (1937) became the last Stevenson lighthouse. The family firm was renamed C & A Stevenson, leaving an ageing and ailing David A to his NLB Engineer's post and with no son to assume his mantle. The situation that had prevailed between his father and uncle was recreated. His nephew D Alan was becoming increasingly frustrated, and impatient in his dealings with the NLB. He had always been rather blunt and outspoken so, when David A finally retired in 1938, the NLB bypassed D Alan and appointed another assistant, John Oswald, in his stead. David A died only a month later.

Eshaness lighthouse, Shetland

In 1940 Charles retired and D Alan became engaged in Admiralty work on the Clyde, ultimately retiring himself in 1952. He had now become the family historian, publishing a book in 1949, *Robert Stevenson's English Lighthouse Tours*, and then another, *The World's Lighthouses before 1820*, ten years later; but his magnum opus on the Stevenson dynasty was never to be finished. Late in 1971, aged 80, he was admitted to hospital and died three weeks later.

Craig Mair reflected on the Stevenson legacy:

Their integrity was unfailing, their busy lives were unpretentious. Their work was creative and important, for even today how many unknown sailors owe them their lives? And yet there is one last, still more revealing, facet of their work. Although they were responsible for so many brilliant inventions, not a single Stevenson ever seems to have taken out a patent. They could have been millionaires but they chose to give their gifts to the world. It was an extraordinary gesture, but then, they were remarkable people.

ten

More Interest and Variety

It seems a curious thing, but natural science is the most popular of all studies ... The marine and birdlife of a lighthouse is full of ever so much more interest and variety than the poor humans who dwell in the round tower ...

Robert Clyne, principal lightkeeper, 1923

Just as the last lighthouses were being automated and the time of the lightkeeper drew to a lamentable close, Ian Cassells, in his entertaining book *No More Paraffin Oilers* (1994), mentioned how, in his experience, he had encountered few keepers who were birdwatchers. Notwithstanding, he was stationed at Rattray Head lighthouse for a time where he would have overlapped with Malcolm Williams, recently posted from the unpopular, and lonely, rock of Dubh Artach.

Whilst I was at Aberdeen University I would visit Rattray with another birdwatching friend Norman Elkins, who was at the time a meteorologist at Dyce Airport and later came to write the definitive text *Weather and Bird Behaviour* (1993) which is now in its third edition. Norman found several rare migrants in Malcolm's garden (including a red-breasted flycatcher) and then enthused him about the art of seawatching. Marooned for a month between high spring tides in the cramped little tower just offshore, Malcolm was soon spending nearly all his spare time logging the movements north and south of all the seabirds that passed by. He would open one of the tiny windows on the seaward side in all weathers and, with hot water bottles strapped round his waist, would stand there for hours, resting his arms against the wall to steady his binoculars, and noting everything he saw. Later he would proudly boast how, in some 200 hours, he logged over 150,000 birds of 33 different species! One of the more memorable days (28 October 1968) some 30,000 kittiwakes were passing the lighthouse each hour. Norman and Malcolm later published a paper on the movements of Manx shearwaters, and some of the data was put to good use in Norman's book. One day however Malcolm looked out a window on the landward side of the tower to discover, much to his dismay, that shearwaters and other seabirds could also pass by undetected behind him!

Later, in the Rattray Archives website, Malcolm confirmed how the lighthouse proved to be a first class observation post for seawatching, while the nearby Loch of Strathbeg (now an RSPB Reserve) and his shore station garden made a welcome landfall for passage migrants every spring and autumn, especially during foggy conditions. A favourite place was 'the run-off pipe from the septic tank where vegetation is lush and insects abound.' During his shore-leave Malcolm also contributed to the first *Bird Atlas*, recording some 55 species nesting in the vicinity of the lighthouse, and the loch, with an impressive list of non-breeding visitors, migrants and vagrants.

Rattray Head lighthouse was completed in 1895 and its very first principal lightkeeper was Robert Clyne, a keen ornithologist. He was brought up on a farm near Kirriemuir in Angus, where his father was a ploughman and friends with a lightkeeper at Scurdie Ness; his older brother John was already a lightkeeper and stationed at Kyleakin. 'On slender threads hangs the course of a lifetime' he would reflect later. As a 19-year-old youth he gave up a job in the local post office to join the Lighthouse Service. After brief inductions at Buchan Ness and Kinnaird Head, Robert began his full-time career at the Isle of May in

Rattray Head in a storm

1878. Being unmarried he had brought his sister Mary to the Isle of May as his housekeeper. It was there she met and, in 1884, married Robert Agnew, a crewman from the NLB tender *Pharos VI*, who then became a lightkeeper himself. Robert Clyne himself married Isabella Davidson from his home village near Montrose. His descendants have carefully constructed the family history, which is available on-line, revealing something of this minor dynasty of lighthouse keepers. One of Robert Clyne's daughters, Lotte, married a lightkeeper and her daughter also married a crewman on the lighthouse supply vessels who was himself the son of a lightkeeper, as had been his uncle, his grandfather and great uncle.

Robert Clyne's new brother-in-law was also called Robert. His father Joseph Agnew was Clyne's principal lightkeeper on May. Joseph was on the Isle of May for 18 years (1868–86) and seems to have been quite outspoken. Agnew is on record complaining – with justification – about the behaviour of day-trippers to the Isle of May, who were frequently drunk and carried guns, a danger not only to themselves but to others. The Commissioners were then minded to prohibit guns on the island. Joseph Agnew proved a competent (and considerate) ornithologist, submitting detailed reports to an investigation run by the British Association on bird migration (see later). Indeed his schedules were deemed the very best of the whole lot.

But back to Agnew's assistant, Robert Clyne. On 28 December 1879, and just over a year into Clyne's term of duty on the Isle of May, a ferocious storm brought down the Tay Railway Bridge; the next day Robert saw the wreckage float by, only to learn of the disaster a few days later. For three of his six winters on the May, storms raged without respite and he witnessed several shipwrecks. Two pilot boats from Newhaven based themselves in the redundant Low Light on the Isle of May. Although 13 October 1881 was ominously calm, the pilots sensed a storm brewing, so the next day they told Clyne they were running for home. They were never seen again. That day, along the south shore of the Firth of Forth as far as Berwickshire, no fewer than 183 fishermen drowned. Twenty-three boats, half of Eyemouth's fleet, were lost, and 129 men (average age 26) perished.

In 1884 Robert and Isabella Clyne moved from the Isle of May to Langness lighthouse on the Isle of Man, until he was promoted at Rattray 11 years later. In 1900 Robert would next be posted to Bell Rock, then in 1908 to the Butt of Lewis, and finally in 1916 to Cromarty, an important naval base. Here the risk of guiding enemy craft meant that the lamp was only lit at the request of the naval authorities; so only a single keeper – Clyne himself – was required.

Although not directly related to our narrative, it might be interesting briefly to digress and consider how the careers of Robert's brother John Clyne and Joseph's son Robert Agnew had developed – fairly typical of NLB lightkeepers at that time. After four years at Kyleakin John was posted to Cape Wrath in 1889, then to St Abb's, Berwickshire, in 1892, Noss Head in Caithness in 1901, Inchkeith in the Firth of Forth five years later, Holy Isle off Arran in 1910 and finally, from 1914 to 1919, to Chanonry Point in the Inner Moray Firth – a one-manned lighthouse since 1900 and where he was close to his brother Robert at Cromarty. John retired to Stranraer after 24 years' service.

Robert Agnew's first posting after he left the *Pharos* a married man in 1884, was to the remotest of all, at Muckle Flugga in Shetland. Two years later he found himself at the Butt of Lewis and in 1892 to the Bell Rock – quite an initiation into the Lighthouse Service! Between 1896 and 1900, he would find Tarbat Ness in Easter Ross a much easier posting, as was his next at Covesea Skerries on the south side of the Moray Firth for four years. But then John was back to islands – Pladda (1904–8), the Monach Isles (1908–10), Hoy High in Orkney (1910–9) and finally Stornoway (Arnish in Lewis?) for three years.

But to return to Robert Clyne. Langness lighthouse at the southeast end of Man had only been established in 1880 but one night all the keepers took ill at Point of Ayre lighthouse, at the opposite end of the island. With no one fit enough to maintain the light Robert Clyne volunteered for the task, but to get there had to drive 30-odd miles over a foggy and dangerous track in a pony and trap. He said it was one of the most fearsome experiences of his career. Nonetheless the Clynes were happy on the Isle of Man and were to remain for 11 years; indeed six of their seven children were born there.

It was his promotion to principal lightkeeper that brought about a transfer to Rattray Head where his youngest daughter was born. Although effectively a rock station, access was tidal so that the keepers were able to transfer ashore – first by boat and latterly by tractor and trailer, to spend their leave with their family in houses nearby. Whilst on the rock and within sight of their homes, they could sometimes communicate using a megaphone, by Morse or semaphore; it would not be until the First World War that a radio was installed.

Some of Robert Clyne's abiding memories of Rattray however, were the many shipwrecks, one of the most tragic being a trawler driven close inshore in a storm. Although he managed to fire a line to attach to the boat, the fierce conditions prevented the keepers and coastguards from operating the breeches buoy. By the time the lifeboat arrived the vessel had been pounded to matchwood and its crew washed away to their deaths.

Robert Clyne served at Rattray Head for four years and nine months, before he received a letter from George Street, notifying him that he was to be transferred to the Bell Rock. Here he remained from 19 April 1900 for seven years and eight months. He would have met John Maclean Campbell, and indeed was to take up his colleague's literary legacy in 1923, by submitting weekly features to Dundee's *The Peoples' Journal*.

It was on the Bell Rock on 25 October 1907 that Clyne saw a strange passerine at the lantern. He went outside to try and catch it for examination but was left only with a handful of tail feathers, which he sent to Eagle Clarke at the Edinburgh Museum. In congratulating the lightkeeper for his actions, Clarke wrote in the *Annals of Scottish Natural History* (1908):

If Mr Clyne was only destined to secure a few of the stranger's plumes, he could not have secured any which in this case, would have revealed the secret of its identification with greater certainty than the pretty black and white feathers which form the tail of this little bird.

Although missing half a tail I suspect the red-breasted flycatcher – for that is what it proved to be – had escaped with its life!

The following year Clyne moved to the Butt of Lewis and again submitted a note to the *Annals* (1909):

> I have had a great deal of station work, and not the time I would have liked for the observation of bird life etc. I have however, considered the place generally rather bare and uninteresting, comparatively; there being no turnips grown in the district, and no cover for small birds. This latter half of October has, however, been exceptional, and I have been pleased to see a few warblers and other woodland birds about for the first time. On 25th and 26th we had blackcaps, restarts and willow warblers. On 1st November I watched for a long time, catching midges on the cliff edge, what I am certain is a red-breasted flycatcher (*Muscipapa parva*), the same species as I got last on the Bell Rock (Annals 1908, p 49). The tail was kept nearly always on the move, and often erected wren-like, and the white feathers, when it made evolutions in the air after insects, were as conspicuous as the white on a wheatear's rump.

In 1915, as he neared the end of his duty there, and clearly not as dismissive of the place as some, he submitted valuable notes on the birds he encountered around the Butt of Lewis, with useful arrival and departure dates. Amongst the rarities he recorded was a squacco heron on 5 June 1913, which stayed for a week. He also recorded the earliest attempts of fulmars to nest on the cliffs of Lewis, noting how they are locally referred to as 'St Kilda gulls' (in Gaelic fulmars are also called *calaman hirteach* or 'St Kilda pigeons'). He noted: 'Had I been a collector and not an observer only, the list would doubtless have had other additions. More especially does this remark apply to the smaller migrants, for of these I have included none whose identification was doubtful.'

Clyne was only one of several lightkeepers encouraged by the editor W Eagle Clark (see next chapter) to submit notes to *Annals of Scottish Natural History* or *The Scottish Naturalist*. This was a time when naturalists still included a shotgun in their standard toolkit and the

Fulmar

maxim 'what's hit is history, what's missed is mystery' was very much the order of the day. Some lightkeepers also resorted to collecting specimens. Whilst on Sule Skerry in May 1900 for instance, James Tomison watched what looked to be a hawk hovering near the tower, and mobbed by 'nearly all the pipits on the island, and as it was within easy range of a gun, it was shot. It turned out to be a cuckoo, the only specimen seen on Sule Skerry.' But, other than slaughtering a sheep for butchering or shooting rabbits, most lightkeepers perhaps would rarely have had need of a firearm for collecting birds, instead able simply to pluck specimens at the glass or pick them up as casualties. Eagle Clarke and his contemporaries carried a gun and only as the 20th century progressed did ornithologists give up firearms to adopt binoculars, notebook and pen as the preferred tools of their trade.

Clyne reflected eloquently:

Seldom is there the chance of diversions that fall to the ordinary. The lightkeeper's cinema is a sky screen, when the storm king plays out his dramas, and the damsel in distress is some frail barque with as pretty a name as ever fiction heroine had.

  Like most keepers, Clyne seems to have been quite a resourceful character for, after his transfer to Langness on the Isle of Man, he took up golf! The keepers had made their own clubs in the station forge, and marked out a course where, on at least one occasion, they played against some professors from a nearby college. He would later make a few holes around the Butt of Lewis lighthouse and kept up his golf after he retired to Montrose. (In 1937 the keepers had a five-hole course on the Flannan Isles. There were also three holes on Hyskeir, off Canna, after the war, and – the most unlikely of all – even a single hole on Muckle Flugga.)

Cromarty lighthouse, Easter Ross

Fishing was a highly popular and practical pastime for nearly all lightkeepers. Robert Clyne was also a keen photographer. When in May 1916, he first went to Cromarty – a strategic naval base – he was in great demand for passport photographs. Being within a restricted area, no one could move without the necessary papers; even the Provost of Cromarty required a passport to visit a dentist in Inverness. Many servicemen departed for battle from Cromarty never to return, and Clyne lamented how his photographs might be the last memory that parents and relatives had of their loved one. Indeed his own eldest son was killed in France in September 1917, only 10 months after enlisting.

Every lighthouse maintained a library of some 400 books, which most keepers devoured eagerly. Clyne writes:

It seems a curious thing but natural science is the most popular of all studies … my own tastes found scope in the libraries for I am deeply interested in bird life.

It was during my early years on the May Island that observations were begun on behalf of a Committee appointed by the British Association to investigate into the migration of birds in connection with meteorological phenomena, and there and then I was initiated into what to me was a service-long study of interest.

The lightkeepers in the service between 1880–1887 sent in several thousands of schedules, a digest from which was prepared for the British Association by Dr Eagle Clark, of Edinburgh. At first, apart from the commoner birds, I knew very little about bird life but at the end of the observation period I was an efficient observer and student. In my time we procured on the May several specimens seldom found in Scotland, and one – the blue-throated warbler [Bluethroat]– on its first appearance in Scotland, and only its fourth in Britain …

Like all his colleagues, Clyne well knew that 'the lighthouse attracts the interesting feathered travellers on their aerial journeys …'. He continues:

Rock lighthouses in particular were in a prime position to note the migration and passage of all kinds of birds and many 'official' observations were carried out for various bodies. Sadly, the hundreds of birds which were closely inspected, were often the casualties which had crashed into the lantern, lured there by its light and warmth, but the casualties were thankfully few in proportion to the hundreds of thousands of visitors. Virtually every type of local sea bird imaginable was to be found at one time or other, on or below their balconies ... we survey the lively scene on the outside grating, the noise of feathered life beating against the lantern, the chirping and screaming and the flutter of wings is incessant ... Thump! Thump! Thump! There they strike. Several have already been killed, and lie inert with legs in the air … The Death Rush. We gather in the slain for inspection. On going out to the balcony, we feel some among our feet, while others, recovering from the shock of collisions, flutter off through the balcony rail, most likely to return and repeat the dose, or perchance meet their death in the mad rush … Some of the birds get tired out and rest fatigued on the grating, but the majority keep up the continual jumping or flying against the panes, or even engaging in a battle royal, the pugnacious little robin, when unceremoniously hustled, showing fight to all and sundry.

Thump! Thump! Thump! There they go again! Ah, I doubt this is a fresh lot of a larger sort, they strike so heavily and to certain death. Must have a look round. Woodcock, indeed, two lie heels up on the grating, the rest must have fallen below or are on the balcony. On opening the balcony door we must be careful or, at the least opportunity, a thrush or starling will fly inside and damage the burners.

Again I gather the dead in, but find no more woodcock. Larks, redwings, and female blackbirds are in the majority this time, but a pair of fieldfares are included in the toll – it is early in the season for them. A few small gold-crested wrens [goldcrests] are also on the lantern now, but their flutters are drowned in the hammering of the larger birds, and they are keeping out of the turmoil in the corners of the angle panes. It is wonderful that such a mite can cross the North Sea.

So the watch flies quickly past!

Next day in the garden and on the adjoining rocks – for we are on a seabound lighthouse – several birds are hopping about unable to fly. The found killed amounts to over 200, few in proportion to the estimated number visiting. Grease, lime, blood and feathers half obscure the lantern panes, and all require liberal cleaning to keep up to inspection order.

Before I leave the light-tower incidents for the larger observations on the rocks and shore I have other things to chronicle. There may be several rushes of migrants during the autumn dependant mainly upon the weather. In the spring migration, the first visitors to the lantern are usually the wheatear chats; then follows the warbler family.

Others of the multifarious families I have seen killed or caught on the lantern are the cuckoo, land and water rails, water-hen, whinchats, flycatchers, and pipits. Of the finch family not many visit, but I have seen the linnet, siskin, the little red-pole, brambling, and chaffinch.

Of seabirds and other residents I have found in the casualty lists are ducks, gulls, sandpipers, lapwing, golden plovers, snipe and curlew. It does not follow that all the birds visiting the lantern are migrants; some may only be attracted to the light on local excursions, but I believe that many of our resident birds are also migratory; for instance, the song thrush, blackbird, and robin. I have never known swallows strike or be attracted to the lantern.

About the most curious visitor I have had in the lantern was one night, when a 'Mother Carey's chicken' or storm petrel, was imprisoned. When caught it vomited an oily discharge, which required several washings and gales of wind and rain to efface from the panes. I took the creature inside for examination, but found it alive with so many parasites, and giving off such an offensive odour, far worse than that from a large cormorant or northern diver, that I was glad to give it freedom …

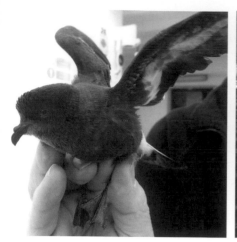

Left: Storm petrel ('Mother Carey's chicken')

Right: Aerial photo of old Sule Skerry lighthouse station

A jack snipe with a broken leg has looked me in by way of the balcony door when the night was calm, and it was open for ventilation. A tiny willow wren has come though the big revolving cowl, and comfortably rested up aloft until the light was put out in the morning. Several times carrier pigeons, with markings on their wings and rings round their legs, have been trapped in the ventilators. A swallow flew in at the kitchen window one morning, and actually rested on the head of the breakfasters seated at the table.

Swifts I have seen quite often at the Bell Rock. One foggy night one came to rest on the kitchen window. At daylight next morning it was very wet, cold and bedraggled, and, instead of flying away, clung, kitten like, to my hand. I took it inside, and placed it in a corner near the fire, where it rested without moving till I took it to the window at 8 o'clock when it took wing, quite refreshed.

Rock life does not begin to stir until March; in the early part of this month there are more of the tiny fry of the billet or saithe to be seen in the pools but no other low-water garbage to attract more than one or two immature gulls occasionally. Things, you see, are inter-dependant. Towards the end of March the first pinkish spawn-clumps of the rock-sucker [lumpsucker], a queer ugly-headed fish, are seen pressed into corners of the rock, with the brilliant rose-coloured male already 'on guard' over his charges. About the middle of April, however, there are still a good many bulky, repulsive-looking females swimming up and down the creeks, apparently looking for suitable corners in which to deposit their spawn …

I have recollections of many pleasant days of spring inducing a scramble on the rocks, covered at this time with a green or maroon-coloured slime and minute white shells. The white dog whelks had not left their winter quarters, but the hermit crabs were early abroad always. Many white and painted top shells hung by yellow fleshy lips from underhanging shelves of rocks, and many cowries were about. The common black winkle and the yellow sea slug lie about. Poking in corners where the dog whelks were piled, many brittle starfish came out of shelter. Rough and exposed as is the Bell Rock, of which I write, the few sheltered parts seem to be veritable haunts of shellfish and crabs.

And sometimes laden winds carry a breath of the country to our sea-girt rock. On breezy days in September thistle down drifts past on fairy wings, and gossamer, sometimes with tiny spiders attached by a faint filament. In flowery June the sweet odours of hawthorn blossom come, like the tinkling of church bells down the wood upon a rustling wind.

It was at Cromarty that Robert Clyne ended his remarkable 43-year career in 1922, aged 63. He then became a custodian in Montrose Museum where he began his series of reminiscences for the *Peoples' Journal*; he died in 1937.

An Orkneyman, James Tomison (1861–1908), whose father had been a lightkeeper, also proved a particularly gifted ornithologist. He was posted as principal to Skerryvore in August 1903. In 1907 – a year before he died – he had published, in the *Annals of Scottish Natural History*, a comprehensive account of the birds he had observed at the lighthouse – no fewer than 64 species. Being no more than 300 ft² in extent at low water and half that at high tide, the basalt rock on which this famous Alan Stevenson lighthouse stands obviously cannot support any breeding birds. A few species can be seen at any time of the year but the bulk that he lists are birds of passage: 'If a large crowd of birds happen to be passing, the scene witnessed from the balcony of the tower is really worthy of being termed one of Nature's wonderful sights.'

Species such as the thrush, pipit, wheatear and starling are to be expected, but Tomison also recorded some choice rarities: a hawfinch and a yellow-browed warbler, both killed at the light; two separate corncrakes; and a Sabine's gull, which briefly one day flew around the tower.

From 1896 Tomison had already spent seven years at Sule Skerry and submitted to the *Annals* a detailed list of the birds he encountered on that tiny scrap of Lewisian gneiss – at 14 ha somewhat larger than Skerryvore, but even more remote:

Situated, as it is, far from the disturbing influence of mankind, these birds have returned to it year after year, finding it a particularly suitable locality for rearing their young. Though the island is now inhabited by the three lightkeepers [sadly the last ones left when the light was automated in 1982], their numbers have in no way decreased, for no one disturbed them; in fact they are looked upon as friends from the outside world, and are carefully protected.

Tomison recorded intimate notes on the breeding birds such as shag, which nested all over the island, often in very accessible situations. On 2 June 1902 heavy seas washed away 200 nests. But the puffin he considered to be:

Left: Puffins with nest material

Right: Black-browed albatross with a Bass Rock gannet in 1968

... *the* bird of the island. To give any idea of their numbers is an impossible task ... They are here, there and everywhere. They first make their appearance early in April, and spend 8 to 12 days at sea before landing, coming close in shore in the forenoon, and disappearing at night. The day they land they begin by flying in clouds round the place, and having made a survey to see that all is right, they begin to drop in hundreds till in half-an-hour every stone and rock is covered.

Arctic terns were the next most numerous. Razorbills, common and black guillemots, kittiwakes, storm petrels and oystercatchers (30–40 pairs) were also breeding. Gannets of course nested on Sule Skerry, 6 km (3 miles) away, although they now breed on Sule Skerry itself. Tomison saw the first Sule Skerry fulmar on 23 February 1901 although 30–40 were already nesting on Hoy in Orkney – the beginnings of their eventual spread from St Kilda to Foula in 1878 and now all round the coast of Britain. Leach's petrels were occasionally captured at the lantern and are now known to breed occasionally on Sule Skerry.

Pairs of rock pipits were permanent residents and when the relief boat *Pole Star* arrived from Stromness: '... crowds of them fly out to meet her, and rest on the deck and rigging. Occasionally one or two take a trip ashore, only leaving the ship when within a mile of the Orkney land.'

A dozen whimbrel with a few corncrake [usually silent]; swallows and skylarks were regular summer visitors but never nested. Tomison's list of rarer migrants seems meagre by comparison with his next posting on Skerryvore, the most significant being a dying greenish warbler on 5 September 1902, which at the time Eagle Clarke decreed to be a first for Scotland, and second for Britain.

Sadly, in 1909 the editors of the *Annals* were to report:

with great regret ... the death of our valued contributor Mr James Tomison, which occurred at the Royal Infirmary, Edinburgh, in September last. Mr Tomison was a keen and excellent observer, and availed himself of the great opportunities afforded him by his calling, as his singularly interesting and well interested papers contributed to the *Annals* on the bird visitors to Sule Skerry and Skerryvore abundantly testify. He was personally known to us [Eagle Clarke had visited him on Sule Skerry in 1904] and won our high esteem, and we regard his death, in the prime of life [aged only 47], as a loss to Scottish Natural History and one which is greatly deplored.

James Tomison's grandparents had been long-lived. John Tomison had died in 1882 aged 84, and his 86-year-old wife Isabella died in 1891, the same year as 'our' James's mother Elizabeth, although she was only 54. 'Our' James Tomison's father – also James – had been keeper at Girdleness lighthouse in Aberdeen when his first child had been born; but the boy died – unnamed – after only two days. 'Our' young James was born in 1861, and his younger brother William two years later, the same year that their father James died, aged only 31. William became a schoolteacher but died at 45, in 1908; it is reported he was buried at sea. That very same year his elder brother, 'our' 47-year-old James also died. All are untimely outcomes for such a notable family.

So, as Agnew, Clyne and Tomison testify, natural history was not an uncommon pursuit for lighthouse keepers. Another keeper, Robert Wilson, contributed to Eagle Clarke's *Scottish Naturalist* (successor to the *Annals),* amongst which was an account of an enormous rush of redwings while he was on Skerryvore. The fall took place almost every night from 17 to 25 October 1914:

What interested me most was the terrible number of birds killed by striking the lantern. The largest number of dead for one night was 131, counted on the morning of 22nd October; these were gathered on the balcony, the narrow strip of grating surrounding the lantern and the gutter round the base of the dome. Undoubtedly, the enormous majority of those killed fell into the sea. We saw dead birds drifting away to leeward, in a steady stream, on which many seagulls fed. We watched this, morning after morning; it was really awful ... There were a great many birds coming, but I have over and over again seen vastly more in the rays of the lantern of the Isle of May, without there being more than half a dozen birds killed.

If it were that the light had the effect of dazzling the birds, and that they struck blindly, the May light should be more fatal than this one, as it is much more powerful and has four flashes to one here, every twenty seconds. From the 17th to the morning of the 23rd this slaughter continued, and on the night of the 23rd the birds arrived as usual, but on the following morning we only found one dead redwing on the balcony. The weather was the same as on the former night, viz. south-east wind, cloudy sky with light haze, the only difference being that the wind was light, instead of fresh. This inclines me to think that the strength of the wind may have something to do with the heaviness of the death-roll. Possibly when a strong wind is blowing, birds coming under the influence of the light only become aware of their danger when it is too late for them to arrest their rapid flight, and in consequence many more are killed than when the wind is light.

In January 1917 keeper William Begg (probably he of that name that Eagle Clarke had met on the Flannans ten years earlier) submitted a largely unremarkable account of autumn migration at Little Ross lighthouse in Kirkcudbrightshire. He noted how, in all his 32 years as a lighthouse keeper, he had never seen swallows at any lantern except here at Little Ross. He also reported a dead wryneck and numerous golden plover and lapwings. Standing out on the balcony one night, a lapwing struck him in the face:

> ... and left me fairly dizzy for some time. After a big migration night I have picked up in the court below and round about the tower as many as thirty six golden plover … the mortality among the lapwings is even greater; we collect them, and have stews and pies … Curlew I have caught in the light room, having on some nights to leave the door open, owing to condensation; these birds walk straight in and make no attempt to fly …

The attitude of lightkeepers, however, was not just death, damage, clean-ups, stews and pies. Many of their accounts, we can see, lament on the death and destruction and also revealed a sympathetic and sentimental side.

In May 1880 for instance, Mr Jack on the Bell Rock saw a kittiwake flying around the tower with a fishing line about a foot long and its handle dangling from its beak. 'I opened the window of my bedroom and leaned out to look at it, when, strange to say, it came up close to me. I took hold of it and brought it inside'. He cut away the line as far down its throat as he could reach. 'Poor little creature, it was a pleasant sight to see it on its flight away, and not as it arrived, weary and worn out.'

This bears an uncanny parallel with a more recent encounter. In early May 1967 the lightkeepers on the Bass Rock spotted an albatross in the gannet colony which was confirmed as a black-browed albatross – the first confirmed record for Scotland. Although gulls frequently mobbed it at rest, the resident gannets showed little aggression and tolerated its presence. Once it was even seen indulging in crude courtship display with a disinterested gannet. As well as consorting with gannets it tolerated the close approach of humans. The lightkeepers observed that it would leave the rock on sunny days to return late in the evening to roost on the Bass. It was known to alight on at least seven different sites on the rock, usually facing out into the wind. But one day the keepers noticed it was in trouble. So in the dark, at 2 am on 19 August, one of the keepers, Duncan Jordan, caught the bird by torchlight. He was able to pick it up without any struggle, to tuck it under his arm and carry it back to the lighthouse where he detached a length of corylene twine entangled around one of its feet.

It was last seen on the Bass Rock on 20 August but on 28 September it was spotted off Fidra. The keepers at Fidra light had watched for it all summer but only had one probable sighting. There had been two previous albatross reports in Scottish waters, only one confirmed. In the 19th century a black-browed albatross lived in a gannet colony on Myggenaes Holm in the Faroes from 1860 until it was shot there in 1894. The Faroese referred to it as 'The King of the Gannets'.

On 20 July 1966 – for one day only – an albatross was seen among nesting gannets on Vestmannsaeyjar off Iceland. It was identified as a black-browed albatross from a photograph but the bird did not reappear at Vestmannsaeyjar in 1967 so it may well have been the one that turned up on the Bass Rock that year. It returned to the Bass Rock for about four months in 1968 and for one month only in 1969. A likely black-brow was seen several times at Hermaness, Unst, close to Muckle Flugga Lighthouse in Shetland in 1970 and what may have been the same bird was positively identified there in 1972. It did not reappear in 1973 but spent most of each subsequent summer from 1974 to 1987, sitting on a ledge amongst the nesting gannets at Hermaness, where it was affectionately known as 'Albert Ross', or just plain 'Albert'. There were no further sightings after 1987, until March 1990, after which it returned annually until last seen on 7 July 1995. If indeed it was the same bird as had been seen on the Bass Rock in 1967 it would have been at least 32 years old when last sighted. The black-browed albatross on the Faroes was seen for 34 years until it was shot in 1894. Had the Bass lightkeeper not gone out to capture the bird and relieve it of a potentially damaging tangle of fishing line, this remarkable odyssey would not have unfolded as it did.

A fifth record for Scotland turned up on Sula Sgeir (a small, uninhabited island in the North Atlantic, north of Lewis) between 25 and 31 August (see Chapter 13).

Peter Anderson and James Ducat (the principal who would disappear on the Flannan Isles in 1901) were two keen naturalists stationed at the Rinns of Islay in the 1880s. On 13 October 1882 the keepers reported:

From midnight to daybreak there was such an enormous number of birds it would have been impossible to calculate their numbers. It was one living stream circling round the light all the time, as far as the eye could see into the fog.

A few nights earlier they had:

... a most extraordinary arrival of small birds ... The island is literally swarming with bramblings, chaffinches, golden-crested wrens [goldcrests] without number, yellow titmice [?], robins, blue janets [dunnocks], siskins, redwings, ring ouzels ... and a little green fellow like a siskin with a brown velvet cap on his head.

One keeper stationed at the Rinns actually wrote to his supervisor in Edinburgh:

I wish you had been here today. You would have had something to look at. I haven't seen the like. In taking the entrails out of a hawk I found in the stomach the bones and feathers undigested of a little crested wren he appears to have swallowed whole

Rinns of Islay lighthouse (a separate island) and cottages on Islay itself

… I could hardly think there were so many birds in all Scotland as we have seen the last few days and nights.

But another supervisor was less interested and reprimanded a rightly indignant keeper for reading in the lantern room – having borrowed a book to help identify the rarer visitors!

The ornithological literature hides many other such references to birds recorded by lightkeepers. Keepers did not seem to have much interest in botany. Nor did they have much experience of insects and other animals. On 7 September 1885 however, Anderson observed:

I may here mention that we have had enormous numbers of what is locally called jenny longlegs above this station for three or four weeks. In the mornings there are great numbers of their legs and wings lying on the pavement. I saw about twenty moss creepers [pipits?] working hard to make their breakfast on them.

On the hazy night of 2 November 1883 a keeper on Fastnet Rock reported a small rush of birds:

… but what was most remarkable was the large quantity of moths, which I could compare to nothing but a heavy fall of snow, they were so numerous.

And in 1886 one of the keepers added:

The time they come is not accurately known, or the direction from whence they come or go – they come in one night and go in the same, and are seen no more.

In his book *The Isle of May* (1960) Joe Eggeling noted how big falls of 'drifted' insects often preceded or coincided with falls of migrant birds. Of the 11 butterflies then recorded on the island, all were known migratory species, in addition, the meadow brown, common blue and small copper have all been captured at light vessels. But, understandably, it is the moths that are attracted to lighthouses at night. Although silver Y moths breed on the island, their migratory movements are often associated with those of painted lady butterflies. He quoted Evans as describing 'rushes' of hundreds of silver Y's between 1908 and 1910 for instance: as a rule males appeared at the lantern in greater numbers than females. The commonest geometrid moths occurring at the light were garden carpets.

In *The Scottish Naturalist* (January–May 1915) mainland entomologist William Evans summarised an extensive list of moths and other insects reported from a dozen Scottish lighthouses, chiefly in the Forth area. Up to the end of 1914 he had received no fewer than 7,500 specimens from the keepers of 241 separate species. These comprised two butterflies and 159 moths, 18 caddisflies and lacewings, 40 Diptera [true flies], 10 beetles and 12 others:

Left: Common blue butterfly

Right: Silver Y moth, pinned specimen

It is usual to think of moths alone as night-fliers, and attracted by lights, but the above shows that not a few other insects have similar habits and are similarly attracted … So far the only specimens that can certainly be regarded as immigrants are the death's head, and the two convolvulus hawk moths.

All the other species are doubtless indigenous in Scotland; but among the examples that visited the lighthouses some may quite well have been of non-local origin. To get to the Isle of May, from which the bulk of the records were obtained, many of the species must have flown across several miles at least of sea, for the food-plants of their larvae do not grow there … A pine weevil and a water-bug (*Corixa*) have been caught beside the lantern, and a new water-catch on the island was soon colonised by four species of the latter genus. Though nothing of outstanding rarity has been obtained, some interesting captures have been made, including several additions to the lists of faunal areas. It is recognised that the North Sea is too wide opposite Scotland for many insects to succeed in crossing it, and that, of course, results such as would doubtless reward on the south-east coasts of England are not to be expected here. Little has so far come from the Bell Rock, but it is evident from what Mr J M Campbell states in his book about that station that insects do occasionally appear there in numbers.

Indeed John Campbell had mused on the phenomenon while stationed on the Bell Rock in the early 1900s.

One of the most interesting insects submitted during Richard Barrington's Irish survey, was a locust (*Locusta cinerascens*) caught at South Arran Island in August 1898. Death's-head hawk moths (*Acherontia atropos*) were received from Coningbeg lightship, 16km

(10 miles) from shore, and about a dozen times from mainland Irish stations. In November 1915 James Wilson on Skerryvore communicated through entomologist William Evans how he found a death's-head hawk moth clinging to the wall of the tower. A strong breeze from the southeast was blowing at the time, and for a fortnight before the wind had been blowing continuously in that direction: '… [this] is another demonstration of the insect's great power of wing. We are left wondering whence it came … '

In 1903 William Eagle Clarke would summarise his insect records from his time on the Kentish Knock lightship (see later) in the December issue of the *Entomologist's Monthly Magazine*.

Lighthouse keepers on Hyskeir (Oigh Sgeir), off Canna

Hyskeir, 8 km (5 miles) west of Canna in the Minch, was owned by Dr John Lorne Campbell (1906–96), not only a renowned Gaelic scholar but also a keen entomologist since boyhood. (Now the island group is the property of the National Trust for Scotland.) He compiled a detailed list of Lepidoptera from Canna (1970) and, over the years he encouraged the lightkeepers on Hyskeir to send him specimens – from what is surely the largest and most expensive moth trap in the world. The island has recently supported colonies of green-veined white and common blue butterflies, while the keepers sent him moth specimens

– of feathered thorn (1949), small dotted buff (1945) and garden tiger. When I visited the lighthouse with John in 1977 he was wielding his butterfly net and sported a tropical pith helmet that he had found washed up on a beach in Canna in 1938 – which just completed the appearance of this fine, old-school naturalist!

Bats are recorded remarkably infrequently at lighthouses, but then these flying mammals are not really known for migratory habits, usually preferring to spend the winter hibernating, the time when aerial insects are scarce. Most species are weak fliers and would be unable to resist anything but the lightest of headwinds. Furthermore, being so small it is likely that few species could lay down enough fat reserves to undertake long, regular migrations. Bats may get blown off course and although they have been recorded at lighthouses it may be that these exhausted wanderers are merely feeding on the insects attracted to the beam. Also it is likely that keepers, concentrating their efforts on migratory birds, did not bother to record bats, which are much harder to see, let alone catch and identify. Barrington specifically mentioned how 'bats have never been reported as striking the lantern.'

From hundreds, if not thousands of submissions from Irish keepers, Barrington was given records of bats at lighthouses '... on several occasions from some of our most out-lying stations (Fastnet, Rockabill, Blackrock [Mayo], and Tearaght), as well as from two light-ships (Lucifer Shoals and Arklow South).'

Three specimens were forwarded to him, and identified as: brown long-eared bat (Tearaght, 4 November 4 1891); pipistrelle or common bat (Arklow South lightship, 21 September 1898); Daubenton's bat (Lucifer Shoals lightship, 24 April 1891). The long-eared and Daubenton's bats were caught alive, the pipistrelle found dead.

All three species are known to travel longer distances from time to time but these may not be migratory movements in the same sense as many birds. Proper migration requires whole populations to move – regularly and seasonally – in and out of an area. The strongest fliers tend to be the ones that are known to fly the longest distances. Only six bat species occur in Scotland. The brown long-eared bat is abundant and widespread throughout Britain except the Northern and Western Isles; one has been reported from Shetland. Noctules, whose range just extends into Scotland, have been seen as far north as Orkney. Natterer's bat tends to be sedentary while Daubenton's and one of our two species of pipistrelle may arrive in odd places. The Nathusius' pipistrelle is a much more migratory species, occurring in Europe but travelling up to 1,900 km (1, 180 miles) between summer and winter quarters. It is only a vagrant to Britain (one found in Jersey had been ringed in Germany) but breeding colonies have recently been found in England and Ireland. The parti-coloured bat occurs in central and eastern Europe and only very rarely in Britain, the most recent in 2001. The very first of these vagrants was found on a North Sea oilrig 270 km (167 miles) offshore in 1965. It is a relatively strong flier and in some parts of its range will fly up to around 1,500 km (930 miles) southwest to hibernation sites.

Lighthouse keepers also regularly observed marine mammals – seals, whales and dolphins. For instance, from the Bell Rock James Campbell recorded several porpoises,

*Right: Lepidopterist John Lorne Campbell with lighthouse
keepers on Hyskeir, off Canna, 1977*

Left: Grey seals ashore on Sule Skerry

a school of 'finner' whales, and a large school of bottlenose whales. Many keepers, I am sure, had one or two interesting encounters with seals, while James Tomison offered more interesting observations from Sule Skerry:

> At one time [it] was the home of thousands of seals, but when man took possession of it they gradually left for other localities where they are less likely to be disturbed. The first winter we were there [1903] from 50 to 60 were still resident … When they were the sole inhabitants they wandered all over the island, and there were distinct tracks in all directions … deep hollows worn down through the grass and even into the ground, leading to and from accessible places from the sea. The female lands in October and November to give birth to her young … When newly born, the young one is covered with a coat of yellowish fur, which in a month or six weeks gives place to the short bristly hair common to the adult. There were several births during my first winter there, and I noticed that the young never attempted to go near the sea until they were a month old … in spite of [the mother's] watchfulness I have frequently got to within a dozen yards of her to observe the feeding operation. She exhibited the greatest symptoms of affection by fondling her offspring with her flippers and keeping up a low crooning sound. Now there are only about a dozen about the island, which land occasionally on outlying rocks.

Although Barrington and the naturalist Alexander More tried to rouse interest in recording other species, bird recording seemed to be the pursuit keepers enjoyed most.

Left: Orcas (killer whales) with calf

Right: Basking shark

With the current interest in whales and dolphins, it is indeed unfortunate that no serious attempt was made to collect and collate lightkeepers' records of sea mammals.

From Hyskeir for instance, the keepers could watch killer whales (orcas) and basking sharks swimming through the narrow channel at the landing stage, and once witnessed an orca hunting seals. I have also seen this at Flannans. Basking sharks were spotted not uncommonly. After the Second World War, Gavin Maxwell (hailed as a conservationist ever since his delightful book about otters, *Ring of Bright Water*, was made into an even more popular film) began a shark fishery off the Hebrides. He based himself on the Isle of Soay, overlooked by the mighty Skye Cuillin and ventured out with a small crew from Mallaig, to kill basking sharks for their livers. His account was published as *Harpoon at a Venture* (1952).

On June 12 [1946] I got a telegram from Uishenish Lighthouse on South Uist: *Good number of sharks here. Davidson.* It was the signal for us to shift our fishing-grounds to the Outer Hebrides … Uishenish, and the bay of which the lighthouse rock forms the south headland, locally called Shepherd's Bight, was the favourite of them all, and we hunted sharks there days without number … The chief lighthouse-keeper, Davidson, became our friend and ally; he kept us constantly supplied with information, and seemed to feel our failures and our successes as keenly as we did …

I made the acquaintance of a friendly and delightful man, and in that and many hours I began to learn about a lighthouse-keeper's life. It was usually a hereditary vocation, he said, a lighthouse-keeper's son often becoming a lighthouse-keeper himself. He told me of the naval standards of efficiency and cleanliness that were traditional; of single rock island lighthouses where a man would do a two-month spell of duty followed by a month ashore; of mad lighthouse-keepers of fiction and of fact; of wrecks and disasters, and of much else …

But then and at all times his first concern seemed to be for our success, and he gave a detailed report of the sharks he had watched during the past few days. They had

Ushenish light-house, South Uist, Outer Hebrides

appeared in the evening only, he said, an hour or two before dusk; first a single fin would show, then another, till within half an hour Loch Skipport and the tide-run off the lighthouse point would be full of them …

… The wind was moderate at last when we sailed for Uishenish on the afternoon of the third day, and when we reached Shepherd's Bight an hour or two before dark it was glass calm, with only the long, slow undulations of a distant ground-swell. It did not take Davidson, gesticulating from the cliff-top, to show us the sharks; they were everywhere, all the way from the jabble of tide-race off the lighthouse point right up into the head of Shepherd's Bight and north to the mouth of Loch Skipport …

We were about half-way between Lochboisdale and Castlebay when I saw a fin a mile ahead of us. Even at that distance it looked huge, and through the glasses I could see that despite its great thickness it had the 'flop-eared' droop at the top which is characteristic of big sharks. As we drew up to him and I could at last see the whole bulk under water, I knew that he was the biggest shark I had ever seen: he was, in fact, the largest that I myself ever saw during the years we hunted them … I was able to fire into him at almost point-blank range, the gun at maximum depression, and the gigantic expanse of his flank practically stationary below me. I had loaded with a slightly increased charge of powder, and I could feel the decks below me shudder with the recoil as the harpoon went squarely home …

The keepers at Neist Point were also helpful to Maxwell in his quest for this second largest fish in the ocean but neither they nor Davidson can be held in any way responsible for the eventual decline in basking sharks in Hebridean waters. In this modern, conservation-conscious era shark hunting is frowned upon and is no longer practised.

In her 2007 book *The Lightkeepers' Menagerie* Elinor De Wire recounted many tales, encounters and anecdotes about animals at lighthouses – mostly pets, and mostly at American lights. At some British stations, and elsewhere, the keepers seem to have been responsible for introducing rabbits to offshore islands – such as the Flannans – offering a welcome source of fresh meat to their restricted diet. There are no rabbits on Shillay where the Monach lighthouse stands, but on the other sandy islands in the archipelago – as elsewhere – rabbits can cause a lot of erosion and are now considered undesirable aliens. Then cats were introduced to the Monachs to try and control the rabbits … beginning to smack of Burl Ives's song 'I Know an Old Woman Who Swallowed a Fly'!

Non-native introductions invariably cause problems, not just to the landscape but especially to vulnerable native species. Nowhere is this better demonstrated than in New Zealand and it is from here that one particular lighthouse-related incident will be forever etched in the annals of conservation. It concerns the global extinction of the flightless Stephens Island wren. As soon as a lighthouse came to be built on the island in 1894 (the optic being supplied from Scotland by the Stevensons) cats were introduced by the lighthouse keepers. Famously – probably unjustly – the blame was laid solely at the door of one called 'Tibbles'. But at least the principal lightkeeper David Lyall had the foresight to send some of his cat's victims to museums (17 in all) and indeed the species came to be named after him – *Xenicus lyalli*. He is on record as lamenting the havoc being caused to the native birds by the cats, and his successor Principal Lighthouse Keeper Robert Cathcart who arrived in 1898, was to shoot over 100 cats in an attempt to remove them. Not only were these tiny wrens so vulnerable by being flightless (one of the few flightless passerines in the world), but they were also so tame and naïve as to fall easy prey to a new, unfamiliar predator. It is perhaps the only instance in the world where a new species was discovered and then brought to extinction (by cats, maybe even a single cat), all in a single year.

Curiously though, there is an old tradition surviving in a few places in Britain and Ireland of 'hunting the wren'. Groups of young men once paraded their kills around their village, hoping that luck, or money, would reward them. And when does this event take place? On 26 December – St Stephen's Day!

eleven

Studies in Bird Migration

No other lighthouse in any country, or of any time, has attained to the same
degree of celebrity as the series of beacons which since the year 1699 have
stood upon those lonely rocks ... [Eddystone] fulfilled beyond all others the
conditions required for the prosecution of the special investigations I was
wishful to carry out ... Only under certain conditions of the weather [do]
the migrants approach the beacon's light and reveal themselves ... I therefore
determined, if possible, to spend a month in such a situation for the purpose of
adding to my personal experience in what has long been a favourite study.

W Eagle Clarke, 1912

We have seen that keepers had been well acquainted with the fatal attraction of lighthouses to migrating birds, probably as soon as the very first powerful lanterns were installed. Ornithologists on the other hand, were slow to wake up to the phenomenon. That is not to say they were not fascinated by bird migration. It was only late in the 19th century that the two began to come together to throw, as it were, some light on the problem.

It was in 1834 that an early call came in Britain for a chain of coastal observers to plot the movement of seabirds. Continental countries like Germany had already begun pursuing such studies, notably on the island of Heligoland. This sandstone island lies 47 km (30 miles) off the German coast and is less than a square mile in extent. Heligoland was occupied by Denmark until the Napoleonic Wars after which, in 1807, it became British. Under such benign rule, the island had become a popular refuge for German writers and artists until Germany resumed it in 1890. Fishing was the major occupation of the islanders but, with Heligoland being on a major migration route for birds crossing the North Sea, they had also long exploited this windfall, together with the local breeding birds, as a source of food. During the wars the population was evacuated and Heligoland was used as a German naval base. It was heavily bombed by the British during the Second World War. Resettled after Germany regained control in 1952, the island is now a popular tourist resort with a resident population of some 1,400 people.

Heinrich Gätke (1814–98), the son of a baker, had studied commerce in Berlin but became a painter instead. It was as an artist that Gätke first visited Heligoland in 1837 until moving there permanently four years later as secretary to the British governor. By 1843 Gätke had taken up ornithology, collecting specimens of rarities for both artistic and scientific purposes. Gätke was to spend 50 years studying bird migration on Heligoland, writing a notable book on the subject just before he died in 1898; his ornithological collection was then acquired by the Prussian Government but was destroyed by bombing in 1944.

A lighthouse had been established on Heligoland by Trinity House in 1811 and replaced under the Prussian administration in 1902 by a brick tower. This was destroyed and its keeper killed by bombing in 1945. In 1947 the Royal Navy detonated 6,700 t of explosive, which totally destroyed the military installations and bunkers sunk into the parent rock of the island. The only structure to survive was a square anti-aircraft tower built in 1941 of reinforced concrete. It was adapted as the new lighthouse in 1952, and faced with a brick veneer some years later, to accommodate a powerful beam visible for 52 km (32 miles). A bird observatory – one of the first in the world – had been established on Heligoland back in 1910 and was formally reopened in 1991; it has recorded over 400 species of birds.

Only slowly during the 19th century did Britain come to realise that there was potential for the study of bird migration around its own shores through the growing chain of lightships and lighthouses. Keepers already knew this of course, but at first ornithologists did not have sufficient confidence that lighthousemen would have the time, the ability, the commitment or the inclination to undertake the data collection required. A Norfolk landowner and naturalist JH Gurney (1848–1922) – whose father of the same name wrote a pioneering monograph on the gannet – tried to inspire one or two keen keepers, but his effort was not sustained.

Of private means, John Alexander Harvie-Brown (1844–1916) – a student of the distinguished ornithologist Professor Alfred Newton's – was extensively travelled at home and abroad. He accumulated a network of zoologically-inclined correspondents, including lightkeepers. In 1874 he had visited Heligoland – the first British ornithologist to do so – and retained a long correspondence and friendship with Gätke. Although also an obsessive collector, Harvie-Brown would be responsible, with various co-authors, for compiling several highly influential *Vertebrate Faunas* of his native Scotland. In 1879 he teamed up with an English naturalist and farmer friend John Cordeaux (1833–99) to formalise a system of recording bird migration by lightkeepers.

In 1879 the pair circulated schedules to about 100 lighthouses throughout Britain: two-thirds of them were returned, duly filled in with bird records. 'So much willing co-operation, we confess, we could hardly have anticipated, especially on a first experiment.' Cordeaux added: 'Great credit is due to the observers of the various stations for the careful manner in which, as a rule, the returns have been made out. Taking them altogether the reports show truthful, accurate, and painstaking observation'.

That year however, the pair reported that:

Left: John Alexander Harvie-Brown, 1844–1916

Right: Starling flock

During long experience at these … stations, the several observers do not remember such great scarcity of birds during the autumn migration. … No rare migrants have been noticed, but this was scarcely to be expected. In this respect the budget of notes supplied by that veteran ornithologist Herr Gätke bears a striking contrast to our east coast reports. On that small island, so favourably suited for observation, Mr Gätke has trained up quite a host of practical observers, and any rare visitant will have to be very sharp if it succeeds in escaping detection.

Many birds are killed at the lantern of the more isolated stations and are blown into the sea. Thus, in 1877, at Skerryvore, in the month of October, the number of birds killed was six hundred, chiefly the common and mountain Thrush [ring ouzel], but including also blackbirds, snipes, larks and one wild duck. The [appropriately named!] Observer, Mr W Crow, was of the opinion that about two hundred more were killed and blown into the sea. They came every night from the 1st to the 6th, at about 8 pm and went away at daylight. I would estimate the number about the light on each of the above nights to be about a thousand.'

The direction of the wind was from south-southeast to south, with haze; and no migration of birds was observed during the day.

On Dubh Artach lighthouse the keeper noted how, 'two hawks are seen every morning while the migration lasts, which come to prey upon small birds resting on the rock.' On Skerryvore another keeper recorded a corncrake on 20 June, an unlikely stopover perhaps for such a secretive creature of the undergrowth with seemingly poor powers of flight.

But already a pattern was beginning to emerge.

Next to the lark, the starling occupies the most prominent position in the reports … A summary of the various returns show four species, larks, starlings, rooks and hooded crows, in the order given respectively, far outnumber any other, and of these four the lark far exceeds the rest in migratory numbers … Migrants have passed the stations at all hours of the day and night, flying at no great altitude and in almost all winds and weather.

The following year (1880) Harvie-Brown and Cordeaux repeated the exercise and were able to add a few new correspondents from the Channel Islands (Casquets and Les Hanois), the Isle of Man, Faroe and Iceland. Reporting on the Scottish results for 1880, Harvie-Brown noted how:

Isle of May stands this year at the head of the list for numerical returns, I having received seven full schedules from Mr Agnew, principally referring to autumn migration. Next comes Bell Rock, but two of three schedules refer to spring migration. Then Sumburgh Head and the Pentland Skerries, about equal … Light-vessels always return the best-filled schedules, and therefore a preponderance of birds are noticed all along the East English coast where light-vessels are most abundant. Whilst upon the East Scottish coast the returns are infinitely smaller where there are no light-vessels, but only lighthouses, whose lanterns are at a greater height.

North Carr lightship, formerly off the Fife coast, at Anstruther

NORTH CARR

In 1881 Mr Knott, keeper at The Skerries, off Anglesey, offered the following observations on fog-signals:

> Terns which breed on the island seem to take no notice of the fog-horn, while others, such as starling, blackbirds, thrushes, larks etc keep off while the horn is sounding, so that very few are seen round the lantern now, while formerly in thick or misty weather during February and November, the lantern gallery would be full of birds; each striking would drop into the gallery and remain till daylight, when, if not too much injured, they would fly; but with strong winds a great number, chiefly starlings, would be killed. It is easy to believe that the hideous sound of the fog-horn, till the birds get used to it, will keep them at a distance … From some lightkeepers I have heard that years ago (the lighthouse then not long erected) the slaughter among birds was much greater than now. Of course the nature of the season would partly account for this, but I think also that the unaccustomed light might attract many a weary wanderer to an untimely death.

To enable an extension of effort, Harvie-Brown and Cordeaux gained support and modest funding in 1881 from the prestigious British Association. A Committee on Migration was established, chaired by the influential ornithologist Professor Alfred Newton (1829–1911), which included Cordeaux (the Secretary) and Harvie-Brown of course with, briefly, a Manx historian Philip Kermode (1855–1932) who would co-ordinate effort in the west of England. Two Irish naturalists, Alexander Goodman More (1830–95) and his protégé Richard Manliffe Barrington (1849–1915), soon brought Ireland into the scheme. More was of independent means but prevented from holding a profession by chronic ill health – so it suited him to pursue a sedentary analysis of the schedules. Barrington graduated in science, then in law, but turned instead to the outdoor life of a land valuer and farmer so that he could pursue his various mountaineering, botanical and ornithological pursuits. He was to prove particularly energetic in his role and kept up the Irish effort until 1900, several years beyond the British Committee's existence.

In 1882 Joseph Agnew on the Isle of May again topped the list, with no fewer than 19 schedules, together with a jar of 43 specimens. Pentland Skerries returned nine schedules, Sumburgh four, Bell Rock three. Harvie-Brown reported how in Scotland:

> … a vast migration took place this year upon our E coast, the heaviest waves breaking upon the entrance to the Firth of Forth, at the Isle of May, and again at Pentland Skerries. Bell Rock also came in for a share, though apparently from the schedules a much smaller one than at the Isle of May. The easterly winds prevailed all along our E coast, generally strong to gales, and the successions of south-easterly and easterly gales in October between the 8th and the 23rd, occurring as they did, just at the usual time of the principal migration, brought vast numbers of land birds to our shores. From Faroe in the north to the extreme south of England this is found to have been the case …

*Far left: Richard
Manliffe
Barrington
1849–1915*

*Left: William
Eagle Clarke
1853–1938*

The winter of 1881/82 had been exceptionally mild and may have resulted in high survival, followed perhaps by a good breeding year, to a truly remarkable fall of goldcrests in particular:

Unquestionably the principal feature of the autumn migrations has been the enormous arrival of the little gold-crested wren. The migrations appear to have covered not only the E coast of England, but to have extended southward to the Channel Islands and northward to the Faroes. On the E coast of England they are recorded at no less than 21 stations, from the Farne Islands to Les Hanois, Guernsey. The earliest notice is August 6th, the latest November 5th, or 92 days; during the same period enormous numbers crossed Heligoland, more especially in Oct, and quite up to the end of the month. On the night from 28th to 29th Mr Gätke remarks: 'we have had a perfect storm of goldcrests, poor little souls, perching on the ledges of the window-panes of the lighthouse, preening their feathers in the glare of the lamps. On the 29th all the island swarmed with them, filling the gardens and over the cliff – hundreds of thousands; by 9 am most of them had passed on again.' Also from 6th to 8th October 'thousands and thousands [of jays] without interruption passing overhead like crows, north and south of the island too, multitudes like a continental stream, all going east to west in a single south-easterly gale.

None were mentioned in the English returns.

In 1882 Philip Kermode, despite repeated requests from the Secretary, failed to submit any reports at all from the west of England. Kermode had turned out, as one historian has commented, 'the lamest possible duck' but his successor to the British Association Committee – recommended by John Cordeaux – proved a key appointment indeed. William Eagle Clarke (1853–1938) was a young civil engineer and surveyor turned museum curator from Leeds. In 1888 he moved to the Royal Scottish Museum in Edinburgh. His friend Philip Manson-Bahr (1959) described him as a handsome man with prominent expressive eyes, sharp cut features, clipped moustache, and carefully parted hair. He spoke with a soft

Edinburgh accent. He made the impression of a successful lawyer or master of business, always appropriately and neatly dressed. He walked well, with measured tread and never openly talked birds with anyone unless urged to do so.'

And thus the massive data collection continued. Harvie-Brown and Cordeaux made heroic efforts to keep the British Association Scheme alive and these had not gone unnoticed. In 1883 they were intrigued to find how:

> Our American friends have made a promising start with similar intentions, but of a much wider scope … Schedules somewhat more elaborate than ours have been issued to lighthouses in America … Some time ago we received application from China for schedules and letters of instruction, but as yet we have had no further communication from that quarter. We would be glad to hear of the scheme being started there also. Most of the lightkeepers there are Scotch and English.

In its eight years the Committee made do with an annual grant of only £175 – with the British Association covering printing costs of the reports – and a further £95 for the final analysis. By 1886 126 lighthouses had been recruited, receiving the standard schedules, instructions and cloth-lined envelopes for the return of legs and wings for further study. It is to the credit of the keepers that they maintained a 60% response. But as time went on the British Association became keen to end the data collection and to see the embarrassing volume of results analysed. Eagle Clarke was charged with this daunting task.

Ornithologists were obsessed with identifying migration routes, but with no clear objective in mind other than data collection. Any conclusions they might infer on the mystery of migration were anecdotal. Undoubtedly though, they added to regional, even national, species lists and highlighted both the frequency and extent of visible passage from lighthouses. Not least they emphasised the true extent of the mortality involved when birds were attracted to the beam. The diligence of the lighthouse keepers – and indeed their accurate, self-taught powers of observation and identification – cannot be praised highly enough. As a sad footnote to their efforts however, I came across an item in Islay's community newspaper (*The Ileach*, 9 September 2007). The distinguished entomologist Miriam Rothschild found the schedules (1880–1887) piled on a lorry ready to be burnt. She managed to save eight bound volumes and letters from the bonfire: 'They deserved a better fate than oblivion'. It would be interesting to know of their whereabouts now, but at least it seems, somewhere they may have been saved for posterity, a tribute to the remarkable band of men who generated them. In their report for 1883, More and Barrington in Ireland had pleaded:

> … whatever results are obtained from this investigation, they will only be arrived at by patiently collecting observations for some years. If the lightkeepers continue to assist us, this can readily be done – without their co-operation annually we are helpless.

Indeed when the British effort finally ceased after 1887, Barrington, at his own expense, continued in Ireland for another decade. 'The Irish Lights Board was favourably disposed to this inquiry from the beginning, provided it did not interfere with the duties of the lightkeepers ... ' He decided to elaborate upon the approach:

The brilliant lights attract thousands upon thousands of winged voyagers, numbers of whom, when bewildered by the glare on dark nights, fly oftentimes against the lanterns with great speed, and are killed striking ... It is hoped that the lightkeepers will not think it too much trouble to cut off and label the wing and leg of every common bird which is killed at their station. Rare or strange birds should be sent entire. These should be forwarded and the costs of so doing will be gladly paid. All species can then be identified with certainty. It is hoped that the notes may extend not only to the night migration, but also have reference to any migratory flocks passing inland, or seen in the vicinity of the station during the day. If no boat comes, steep birds in methylated spirits – it does not injure them.

Over 16 years (until 1897) keepers from over 50 lighthouses and lightships around Ireland supplied him with 39,000 records of 170 species: 3,000 specimens yielded 1,600 wing measurements. Referred to as Barrington's 'avian herbarium' he was to summarise its results in the year 1900, as a 1,000 page volume entitled *The Migration of Birds as Observed at Irish Lighthouses and Lightships,* of which only 350 copies were printed (thus they can now fetch a couple of thousand pounds each on the antiquarian market!). Barrington gratefully acknowledged how ' ... the time and attention of the lightkeepers were given voluntarily and without payment.' His collection is lodged in the National Museum of Ireland (Natural History), where it is available for research; indeed currently it is part of a study of climate change.

Richard Manliffe Barrington led a very full life. In his youth he had gone mountain climbing in the Alps and Canada. In 1883 he visited St Kilda where, with two islanders, he became the only outsider to scale the notorious 73 m (240 ft) Stac Biorach. He was a skilled botanist, as well as ornithologist – 'an ideal companion, full of enterprise, originality, humour and a never-failing friendliness'. Barrington returned to St Kilda in 1896 having just failed to achieve a landing on Rockall with the Irish botanist Robert Lloyd Praeger (1865–1953) and the Scottish naturalist Harvie-Brown. At the age of 66 Barrington died suddenly on 15 September 1915 as he was motoring from Dublin to his home in Bray, County Wicklow.

Back in Britain, when the final reports were published Eagle Clarke felt there were still significant gaps. So he embarked on a series of visits to remote lighthouses, beginning in September 1898 with a rather disastrous visit to the island of Ushant, 20 km (13 miles) off Brest and the most northwesterly point of France. Only 15 km^2 (5.8 sq.mls) in extent and with a population of around 2,000, the low-lying granite island had two lighthouses, one in the northeast and the other in the southwest. The keepers proved helpful enough and reported up to 500–600 birds killed at the lanterns in some autumns, especially starlings. Indeed 10 years previously they had recorded as many as 1,500 dead birds. Although they were beginning to make some interesting observations, the brilliant weather yielded little in

Eddystone lighthouse

the way of bird movements for Clarke and his companion TG Laidlaw. Furthermore, at the time, Anglo-French relations were somewhat strained and the gendarme sergeant from the mainland harried the British pair so much that they had to leave after only six days. Before returning home they spent a useful week on Alderney in the Channel Islands.

Undeterred, Eagle Clarke was next able to arrange a month living with the lighthouse keepers on the notorious Eddystone.

> No other lighthouse in any country, or of any time, has attained to the same degree of celebrity as the series of beacons which since the year 1699 have stood upon those lonely rocks ... [Eddystone] fulfilled beyond all others the conditions required for the prosecution of the special investigations I was wishful to carry out ... Only under certain conditions of the weather [do] the migrants approach the beacon's light and reveal themselves ... I therefore determined, if possible, to spend a month in such a situation for the purpose of adding to my personal experience in what has long been a favourite study.

Eagle Clarke deemed autumn most suited and the south coast of England best placed to witness the departure movements from Britain. 'An ideal watch-tower would be one situated well out in the waters of the English Channel ... The famous Eddystone lighthouse offered all these advantages.'

With a strong recommendation from the Royal Society, Professor Alfred Newton and Sir Michael Foster forwarded an application to Trinity House, which was 'most graciously granted by the Elder Brethren.' Eagle Clarke was able to take up residence with the lightkeepers on 18 September 1901.

Landing on the rock is somewhat exciting work for a novice, and is effected from a surf-boat towed out by the relief steamer for the purpose. This boat approaches the rock at low water, and anchors as near the base of the tower as the surf which eddies around it will permit. Communication with the lighthouse is then established by means of a rope. To this rope the person about to land clings with his hands, and places one of his feet in a loop formed for that purpose. In this dangling fashion the intervening waters are crossed, the rope being paid out from the boat, and hauled in by the winch placed halfway up the tower ...

Life on a rock station has, of course, its little trials. He who would dwell there must, among other things, be prepared to share in all respects the lot of the keepers, their rations, and their dormitory. He must also be content to be shut off from communications with the outer world until the monthly 'relief' comes round, when, weather permitting, his incarceration ends and he returns to the ordinary comforts of everyday life. There was one feature in the life on the Eddystone which was decidedly trying to an amateur, namely, the firing, every three minutes during fog or haze, of a charge of tonite, an explosive producing a terrific report which can be heard some 15 miles (24 km) or more. The keepers were able to sleep peacefully during these operations – an accomplishment I did not succeed in acquiring. I may say at once, however, that the novelty of the situation, the interesting nature of my self-imposed work, and last, but not least, the great kindness of the keepers, far outbalanced those trivial discomforts which are inseparable from such a life; and I shall ever look back upon my sojourn in that lonely observatory with extreme pleasure and satisfaction ...

My visit included a period when the nights were brilliantly moonlit and cloudless, during which, no doubt, great passage movements were performed. When such conditions prevail the birdwatcher may rest in peace in his bunk ... Gales were not infrequent and arrested the emigratory movements.'

Those that Eagle Clarke witnessed came to be 'ineffaceably impressed' on his memory. His first took place between 3 am and 5 am on the 23rd September, during a 40 to 48 mph, south-easterly gale. As soon as the wind moderated, and despite the heavy rain, migrants began to appear in numbers.

I was aroused from my sleep by the keeper on duty with the words 'Birds, sir', and was on the gallery a few moments later ... The birds were flying around on all sides, and those illumined by the slowly revolving beams form the lantern, had the appearance of brilliant glittering objects ... I was not a little surprised to discover just how extremely difficult it was to identify [them] ... the smaller species had to be lifted from the lantern ere their identity could be ascertained; and the birds careering around became mere apparitions on passing from the rays into the semi-darkness beyond. A number of species undoubtedly escaped detection; but the following are known to have participated in the movement, those marked with an asterisk having been either killed or captured:

– song thrushes*, redstarts*, sedge warblers*, pied flycatchers*, yellow wagtails, turtle doves, redshanks and curlews. The song thrushes, yellow wagtails and turtle doves were most in evidence. The turtle doves often approached the lantern, yet they recovered themselves sufficiently to avoid striking. The yellow wagtails killed included both adults and young … The rush was evidently composed of departing British summer visitors, spurred to move southwards by the very unsettled weather of the previous few days.

Birds around Eddystone lantern (Marion Clarke)

There were several other minor rushes but the first northern migrants, such as redwings, appeared after a gale lasting three-and-a-half days: at 2 am on 10 October in a gentle northwesterly breeze, as soon as the clear and starlit sky became overcast. But the chief movement Eagle Clarke witnessed was from 7.15 pm on the night of 12 October, continuing until 5.45 am. The first birds to appear, from 7.30 pm, were a few starlings, which persisted throughout the night. They were followed by blackbirds, skylarks, stonechats, redwings, fieldfares, wheatears and song thrushes, arriving and passing on in a steady stream, with many striking the lantern. This rush intensified after midnight with the addition of chaffinches, grey wagtails, goldcrests, fieldfares, white wagtails, meadow pipits and curlews.

A fresh rush at 5 am included a grasshopper warbler while a party of herons passed close to the dome, calling loudly.

> It would have been possible to have captured some of them in great numbers; and, as it was, the killed or injured which did not fall overboard, included 76 skylarks, 53 starlings, 17 blackbirds, 9 song thrushes, and examples of the redwing, mistle thrush, stonechat, chaffinch, meadow pipit, grey wagtail, white wagtail, goldcrest and grasshopper warbler.

By daybreak all but a few starlings resting in the recesses of the windows had departed. Eagle Clarke considered the whole episode beyond his powers of word-painting.

The last, and most instructive, falls of migrants took place on the night of 15 October and continued until daybreak to afford one a clear and unmistakable demonstration of the influence of weather. The sky was bright and starlit but with regular intervals of cloud and haze and drizzle, during which the beams grew most conspicuous. It was then that the birds approached the beam only to disperse again when the sky cleared. The usual species were involved, with the addition of wrens and storm petrels; a number of unidentified waders passed overhead. Song thrushes and skylarks were the most abundant, but starlings remarkably scarce.

Eagle Clarke also observed daytime movements at sea, which included regular passages of meadow pipits and wagtails, usually soon after dawn with almost nothing to be seen in the afternoons. Swallows were observed flying south on seven days, but few groups of waders were seen – the most notable were two red-necked phalaropes during periods of unsettled weather; both sought shelter at the foot of the lighthouse where they were 'assiduously and unceasingly engaged in the capture of some minute surface-swimming creatures, probably crustaceans.' Sometimes storm petrels were similarly engaged. The first purple sandpipers were seen on the exposed rock on 11 October and a few seem to be regular in late autumn and winter. Amongst the seabirds he recorded was a rare Sabine's gull, moulting from adult into winter plumage. Great shearwaters were common, sometimes harassed by great, pomarine and Arctic skuas. Manx shearwaters were less common and sooty shearwaters rare. Gulls and gannets were regularly observed of course, but few terns.

In the 32 days he spent at the lighthouse, only 14 were entirely unsuited for migration, owing to adverse weather conditions. Eagle Clarke's enthusiasm obviously rubbed off on his hosts, for the keepers continued to send him completed schedules and wing specimens for three further years. As a result he compiled a list of 75 species seen at Eddystone.

Having now observed a southerly migration, two years later Eagle Clarke endured a month, on the Kentish Knock lightship (32 km (19 miles) off Margate) to experience east to west movements. Permissions were again granted and he joined the ship on 17 September 1903 where he remained until 18 October: 'Life in a lightship (apart from *mal de mer*, from which even the crew are not on all occasions immune) is undoubtedly one of considerable hardship and discomfort.' Off the Kentish Knock sands the strong tides made the ship ride broadside to easterly and westerly gales, confining Clarke to his bunk consoled in the knowledge that bird migration was impossible anyway. Nonetheless he claims to have enjoyed the best of health whilst on board, only the relentless siren 'of exceptional power' in times of fog or haze rendered sleep impossible. Remarkably though, he did get used to it.

The ship's white revolving beam seemed especially attractive to migrants passing in the night and he was to find a cross migration 'of a very varied and complex nature, in which respect it is probably not surpassed by any other station on or off the British coasts.' There were summer migrants moving south for the winter from northern Britain as well as North and Central Europe: 'Not a few alighted on the ship, where some of them being both tired and hungry, spent a considerable time resting or busying themselves in an active search for insects, of which we had numbers on board at the time.'

Left: Immature wheatear

Right: Song Thrush

He would publish his insect observations in the December issue of the *Entomologist's Monthly Magazine*. All the grounded migrant birds, when they finally departed, headed for the coast of Kent. Perched on the rail or rigging, he also spotted the occasional rarity such as an icterine warbler and a blue-headed wagtail. Clarke soon found that having an unfortunate sailor stationed on the sloping roof of the lantern, armed with an angler's landing net, enabled many birds, ranging from a goldcrest to a rook, to be captured for the purpose of identification!

He was convinced that, contrary to popular opinion, the birds that appeared at a lantern were not lost and making their way to the light in the absence of any other directive impulse. They were positively attracted to the light.

At Eddystone, the emigrants which I saw in such numbers had practically only just left the land behind them, and had not had time to get lost when they appeared at the lantern. Another important fact in support of my contention is that the birds never appear at the light-stations at night, except when the rays are remarkable for their luminosity … Another significant fact is that they do not seek stations having red or green lights. Such lanterns, I am informed by the keepers, are seldom if ever visited under any conditions … When the Galloper lightship [east of the Kentish Knock] had white lights, great numbers of birds were allured to its lanterns, but now that the light is red, bird visitors are almost unknown. If the birds were lost, why should they seek a white light, and avoid one that is red or green? That the migrants may and do become confused, and for a time perhaps, lost after the excitement and fatigue occasioned by their attendance upon the lantern, I can well imagine.

Another fact that intrigued Eagle Clarke was that the vast majority of birds attracted to the Eddystone and Kentish Knock beams were passerines – with the exception only of a storm petrel and a kestrel. Notwithstanding, he had heard waders, herons and other birds passing on by, unconcerned and invisible. But he could not think of any reason why songbirds should be so vulnerable.

The night movements he witnessed on the Kentish Knock were from an entirely new standpoint – namely from below:

From the deck of a lightship, one realised more fully the terrible loss of life that is involved by these nights at the lantern. Here one saw birds actually falling thickly around, and even heard them dropping on to the surface of the water. Such scenes often lasted for hours – ten and a half hours on the 17th to 18th October – and the sacrifice of life on this and other occasions was simply appalling. Some of the victims, indeed the majority, were only stunned or slightly injured, and thus met with a miserable death at sea.

He was also moved by a kestrel that appeared at 8 pm on 18 September and careered around without a break or rest of any kind until 1.30 am the next day.

Some movements, by day or night, were from the Kentish coast towards Belgium and included house martins, meadow pipits and pied wagtails, with some wheatears, skylarks and starlings. All these species, together with a few whitethroats, spotted flycatchers, thrushes and blackbirds, were at times greatly attracted to the light beam. Eagle Clarke found the east to west flights the most interesting of all. He knew that many of these migrants would settle down to winter in Britain, others moved on to winter in Ireland while yet more moved south across the Channel. These included flocks of skylarks, chaffinches, with small parties of tree sparrows, meadow pipits and starlings, even some waders such as ringed plover and lapwings.

A 10°C drop in temperature on 10 October would prompt the greatest diurnal movement of birds that Eagle Clarke had ever witnessed. It began at 8 am and intensified during heavy rain from midday until 4 pm, with continuous flocks of skylarks, tree sparrows, chaffinches and starlings all heading towards England from the continent. From a distance the latter, flying low over the sea, resembled dark clouds, which contrasted with the leaden, white-crested billows. Indeed, so torrential did the squalls become in the afternoon, that:

> ... it was necessary to sound the fog-horn, whose hideous yells added a weird accompaniment quite in harmony with a scene which, apart from its intense interest to the naturalist, was dismal and depressing in the extreme ... How the migrants braved such a passage was truly surprising. How they escaped becoming waterlogged in such a deluge of wind-driven rain was a mystery. Yet on they sped, hour after hour, never deviating for a moment from their course, and hugging the very surface of the waves, as if to avoid as much as possible the effects of the high beam wind. It was surely migration under the maximum discomfort and hardship, indeed under conditions that approached the very verge of disaster for the voyagers.

Smaller northward and westward movements were recorded a few days later, but this time with skylarks and chaffinches passing flocks of the same species heading south across the Channel. On 17 October the first – overdue – small flocks of rooks and jackdaws appeared heading west. The following day Eagle Clarke himself departed west to the Thames estuary where he disembarked the Trinity yacht *Irene* at Southend. During the four-and-a-half hour voyage they crossed or ran parallel with continuous flocks of starlings and skylarks in their tens of thousands, with some chaffinches and tree sparrows, all heading westward just above the surface of the calmest of seas and in the finest of weather. In his month at sea he had recorded 59 species on or from the lightship, and was able to extend his experience considerably, enough to begin formulating ideas and theories on bird migration.

Starling flock

Left: Leach's petrel

Right: Young peregrine falcon

These ornithological excursions were taken during Eagle Clarke's annual leave so hardly served as relaxing holidays. With Laidlaw again, he spent 16 days on the Flannan Isles in 1904. Impressed with the migration schedules that the keepers returned, and with the sanction of the NLB, they were able to stay at the lighthouse, built in 1899. Thirty-two kilometres (19 miles) west of Lewis, the islands are sometimes referred to as The Seven Hunters, the largest – Eilean Mor – only 6 ha in extent. Its lighthouse stands on the highest point 85 m (280 ft) above the sea. The pair were slightly frustrated at being so restricted, wishing they could extend their researches to some of the smaller, satellite islands. Cliff-bound, the islands support respectable seabird colonies in summer, notably puffins and Leach's petrel; the fulmar then only recently establishing on the other islands, with two pairs nesting on Eilean Mor for the very first time recorded in 1904.

After a rather rough passage in the lighthouse tender *Pole Star*, Clarke and Laidlaw arrived at Eilean Mor on the morning of 6 September, to remain until the next relief on 21 September. Both possible landing stages are impressive indeed (and it was there at the west landing that the three keepers were dashed to their deaths in 1900).

Although the Flannans' revolving beam was a powerful one, few birds visited it during their stay because of the clear, bright night conditions. Furthermore a young peregrine falcon:

... proved a scourge to all the small migratory birds resorting to Eilean Mor. It used to dash many times a day over the exposed plateau in search of prey, making sad havoc in the ranks of the travellers. All our efforts to put an end to its ravenings were futile, owing to the impetuosity of its dash and suddenness of its appearance.

Presumably Eagle Clarke was trying to shoot it!

One remarkable movement the naturalists witnessed was the arrival of jack snipe on 16/17 September, dawn breaking to strong northwesterly winds and heavy rain.

They were in astonishing numbers, and sheltering behind rocks, tufts of grass, in the small pools and runnels, and even down the face of cliffs on the north side. In walking across the island I put up a continuous stream of them, in spite of the fact that the birds sat like stones, and only those arose on the wing which lay directly in my course, and when I was close upon them … I have no doubt they were abundant on the other islands of the group, and probably especially so on the adjacent and comparatively flat-topped Eilean Tigh. The birds were present in numbers the entire day, but nearly all disappeared during the night.

Clarke and Laidlaw found many storm petrels nesting in the ruins of the old buildings, in holes in turf, and under stones among grass. Some chicks had only recently hatched ('tiny balls of pretty lavender-grey down'); others were fully feathered with only tufts of down on their bellies. The adults only came in from sea at night and occasionally visited the lantern. Interestingly they found the otherwise rare Leach's petrels more abundant than storm petrels at the Flannans. They laid their single egg earlier – the keepers finding the first on 29 May – but otherwise the habits of the two species seemed similar. Rather surprisingly the keepers maintained that Leach's petrels would capture moths at the lantern. Eagle Clarke – or rather the lightkeepers – proved that this species nested on the island, the only other two known breeding sites in Britain at the time being the equally remote islands of North Rona and St Kilda.

Two of the Flannan lightkeepers, Begg and Anderson, took a keen interest in the bird activity and continued to supply Eagle Clarke with records up until their own departure in 1910. The Flannans' list was now standing at 115 species, including some nice rarities, and quite a respectable total considering the small size of the islands and with Eilean Mor being the only one accessible to them.

On the return voyage Eagle Clarke was delighted to discover that the *Pole Star* was to take stores to Sule Skerry – the most remote lighthouse in Britain – before returning to Stromness. Less than 1 km in length, this lump of gneiss is only 14 ha (34 acres) in extent and lies 35 km (21 miles) northeast of Cape Wrath. Its highest point is merely 14 m (46 ft) above the sea, where the highly exposed lighthouse was built in 1895. During his few hours ashore Clarke noted 13 species of migrants, including several Lapland buntings, although the overall total for the island actually stood at a respectable 103, including Siberian chiffchaff and northern willow warbler. He was puzzled to discover from the keepers that no more than 12 starlings have ever been recorded on the little island. Moore, one of the keepers, once

Fair Isle from the north

found several dead crossbills in the Arctic tern colony and reckoned that the little migrants had been killed by the angry terns. Lightkeepers' observations on Sule Skerry dated from 1899 and one of the most active was our friend James Tomison – 'a most painstaking and capable observer'. He was to spend seven years as assistant lightkeeper on Sule Skerry, before going on to become principal on Skerryvore from 1903 until 1908. He published papers on the birds of both lighthouses in 1904 and 1907 but, sadly, by the time Clarke's monumental *Studies in Bird Migration* was published in 1912 his good friend Tomison had died.

Eagle Clarke spent a further 14 weeks on other isolated islands such as St Kilda, 72 km (45 miles) west of the Outer Hebrides, and Fair Isle, situated strategically halfway between Orkney and Shetland. In all he would clock up 47 weeks' residence in his pursuit of migrating birds. But it was in 1905, on the first of seven subsequent visits, that he quickly recognised Fair Isle's enormous potential. He came to call it 'the British Heligoland.' Fair Isle (part of Shetland, lying 40 km (25 miles) to the north) then had a population of 130, mostly residing at the south end. It is 4.8 km (3 miles) long and 2.4 km (1½ miles) wide, the highest point being 217 m (712 ft) above sea level and almost entirely surrounded by high sandstone cliffs. In 1891 a lighthouse was built at each end, both becoming operational a year later. It was at the South Light that Eagle Clarke usually stayed.

'On consulting a map of Scotland with a view to selecting a bird-watching station in which to spend my autumn vacation in 1905 I was much impressed with the favourable situation on Fair Isle'. His companion this time was young Norman B Kinnear (1882–1957) from Edinburgh, who would later become the Director of the British Museum (Natural History). In the South Light, where according to one islander they were 'living on bad food and good whisky for a fortnight', they fell in with a 16-year-old, called George Wilson Stout of Busta (1888–1916), who possessed a remarkable knowledge of the island's birds. Eagle Clarke provided him with books and field glasses and, after a few weeks' training, came to regard the youth as the official Fair Isle bird recorder. Young George left the island in 1909 when Clarke found him employment with a Glasgow taxidermist. But the museum curator and his protégé returned to Fair Isle in 1909 and 1910, and also visited St Kilda in 1910, Auskerry in 1913 and finally the Butt of Lewis in 1914. Sadly George Stout died at the Somme with the Army Medical Corps in 1919. His mentor Eagle Clarke wrote of him: 'He was a man of irreproachable character, amiable in disposition, and indefatigable in the pursuit of nature knowledge. His loss at the early age of twenty-seven will be greatly deplored by all who knew him.' Back on Fair Isle he was replaced by his cousin Jerome Wilson of Springfield (1880–1948). Wilson joined the navy in 1915 but survived the Great War to return home and resume his island life.

Left: George Wilson Stout of Busta, Fair Isle 1888–1916

Centre: Kenneth Williamson 1914–77

Right: W Eagle Clarke in later life

It was only in 1910, after the first of several more visits to St Kilda, that Eagle Clarke felt himself sufficiently qualified to write a book about bird migration, and his monumental two volume work *Studies in Bird Migration* was published in 1912 – all 670 pages of it. Nearly a half of the second volume is devoted to Fair Isle alone.

Like his colleagues on the British Association Committee, Eagle Clarke was convinced that migrating birds followed well-defined trackways through northern Europe, and firmly promoted his belief that the British Isles lay on such a trunk route linking Norway with France and the Iberian Peninsula – a route along which millions of birds travelled unerringly, season after season, regardless of the prevailing wind.

An alternative theory of drift migration would be proposed in 1918 by the Misses EV Baxter and LJ Rintoul (see later) as a result of their considerable experience on the Isle of May and – with the benefit of bird ringing – famously developed in 1952 by Ken Williamson (1914–77), the first warden of Fair Isle Bird Observatory. They postulated instead that major influxes of migrants in Britain were entirely fortuitous, only when displaced from their normal continental route by easterly winds. Modern opinion from radar and other studies indicate that the situation is more complex, influenced by a range of factors such as cloud structure, wind direction and wind strength together with the height of bird flight and when they embark on their journey. Ideal autumn conditions are within, or to the east of, an anticyclone where winds are light and variable: frequent breaks in the cloud facilitate navigation reducing the risk of drift. The longer birds delay their departure the more likely they are to encounter unfavourable weather. Under the powerful urge to breed, there is greater urgency to migrate in spring. But with more likelihood of bad weather at the end of the journey north, too early a departure in spring poses greater risk. This whole debate however, would not have progressed in the first place without the early ideas raised by Eagle Clarke.

There is no doubt that Fair Isle is exceptional for its migrant birds, adding (up to 2010) 27 vagrants to the British list and bringing the island's own total to a handsome 374. Between 1905 and 1911 Eagle Clarke recorded 207 bird species, more than half the British list at that time. His final visit to Fair Isle was in 1921 aged 68, accompanied by Greenock-born Surgeon Rear Admiral James Hutton Stenhouse (1865–1931) who subsequently took over the Fair Isle bird studies. On that occasion the pair stayed at Pund, a crofthouse then owned by Mary, Duchess of Bedford (1865–1937), a respected ornithologist of private means. Eagle Clarke was remembered on the island as 'a very nice man who was kind to the bairns'; he visited the school where he enlisted the young pupils to collect beetles for him (Dr Ian Pennie, 1987).

Although his passion was bird migration, Eagle Clarke published on many aspects of vertebrate biology and taxonomy. For example, in 1881 he was co-author of *A Handbook of the Vertebrate Fauna of Yorkshire* and, in 1907 of *The Birds of Yorkshire*. With the explorer William Speirs Bruce (1867–1921) he wrote up the ornithological results of Bruce's Scottish National Antarctic Expedition. A British subspecies of song thrush was named after Eagle Clarke, while he himself first described the Hebridean subspecies of wren. He also clarified the subspecific status of the St Kilda wren and, supplying the local lads with traps, was able finally to do the same for both subspecies of St Kilda mice, barely two decades before the St Kilda house mouse went extinct.

By 1891 Eagle Clarke had become editor of the journal *The Scottish Naturalist*, and joint executive editor (with Harvie-Brown) of its successor *The Annals of Scottish Natural History*. (The latter reverted to *The Scottish Naturalist* in 1912, again with Eagle Clarke as editor, until its demise in 1921.) Clarke encouraged many of his lighthouse keeper friends to submit natural history notes and articles. He became President of the British Ornithologists' Union in 1918, and four years later he was awarded the Union's Salvin-Godman Medal. He was appointed an Honorary Fellow of the American Ornithologists' Union, and was awarded a Doctorate from St Andrew's University. One eminent ornithologist considered how Eagle Clarke's 'importance as a leader in every branch of Scottish ornithology was particularly outstanding' (David Lack, 1959). On his retiral from the Museum in 1921 Eagle Clarke was appointed honorary curator of its bird collection. Latterly however, as Philip Manson-Bahr concluded, 'his brain became clouded and he was no longer conscious of the world around him': he died in 1938.

twelve

Bird Strikes and Observatories

*These members of the spring migratory movement often come to grief on our
lantern, and when one considers the number of lighthouses round our coasts, it will
be understood that the death-toll from this cause alone must be extremely high.
Designed to save life, we unwittingly lure our feathered friends to their destruction.
… A new light is being completed on the Bass Rock, and on the first of December,
yet another factor in our dwindling list of visitors will be in operation – ostensibly a
lighthouse – but to our feathered friends, alas! a veritable slaughter-house.*

John Maclean Campbell, 1902

In September 1956 three young bird watchers – Tony Marr and two friends – went to camp
at St Catherine's Point, on the southern tip of the Isle of Wight. They gained permission
from the keepers to go up the lighthouse on the nights of 8th/9th and 9th/10th to record
any migrants attracted to the light. At the time St Catherine's had the most powerful light
(now the third most powerful) in the British Isles – 5.25 million candlepower visible on
clear nights for up to 48 km (30 miles). On the first night there was a strong to gale force
east northeast wind blowing with occasional heavy rain and it was not until well after 2 am
that the first birds appeared – turtle doves, several of which struck with one fatality (later,
the boys cooked it for lunch) and some passerines. A big movement –mainly passerines –
began after 5am, which continued until dawn: *c.*150 turtle doves, *c.*40 pied flycatchers, 50–
60 redstarts, 30 wheatear, *c.*25 whinchat, and assorted warblers. Hundreds more birds were
seen too far away to be identified. Perches that the RSPB had erected around the light tower
effectively reduced fatalities. After daybreak the bushes around the light were full of birds.

The winds on the next night were less strong but the same direction, with no rain but
occasional fog. The bird activity was in full swing by 1 am, and again continued until dawn
and involved 'incredible numbers'. Tens of thousands must have passed and the boys recorded
hundreds of wheatears and willow/chiffs (it is sometimes difficult to tell willow warblers
and chiffchaffs apart), over 100 common whitethroat, *c.*80 pied flycatchers, *c.*40 redstarts,
24 sedge warblers (five killed), seven garden warblers (one killed), two spotted flycatchers,

St Catherine's Point lighthouse at sunset

two grasshopper warblers, a goldcrest, a robin, a nightingale with a few whinchats, lesser whitethroats and chiffchaff. There were only three turtle doves this time, with some common sandpipers, five dunlin, a redshank and a lapwing, but the two highlights were an ortolan bunting and an aquatic warbler. About 10 birds were killed against the glass and once again, the next day, the bushes were full of birds, including three blackcaps, a red-backed shrike, a hoopoe and another aquatic warbler – in all a decent tally. Relatively few birds were killed in Tony Marr's two nights at the St Catherine's lantern (*Junior Bird Watcher*, 1957).

Many years later, Peter Hill – art graduate, poet, author, art critic and, briefly in his youth, supernumerary lightkeeper – published a lightkeeper's account of a significant bird strike in his eloquent and entertaining book *Stargazing* (2003). In 1973 he spent about six months experiencing the life of a supernumerary keeper, during which he was posted to several different Scottish lighthouses – Pladda, Ailsa Craig and the Hyskeir. It was during his last night on Hyskeir (off Canna) in October that he experienced a rush of migrants attracted to the lantern, and he adds a modern perspective to the depressing events. He was first alerted by a constant tapping on the windows. 'It's the birds,' his colleague intimated, 'they've arrived.' Hill had never seen anything like it before, 'Feathers and beaks and eyes and claws seemed to carpet the window with a double layer of darkness.' The two keepers went outside, careful not to let any of their visitors over the threshold:

> I looked upwards. There were thousands and thousands of birds, circling in the beam of the turning light. Even from the ground I could tell they were of different sizes, different types, specks of wren and blobs of thrush and redwing. But they glided round so carefully, as if on a carousel … Round and round they went, following the beams of light. But there were thousands more … For, as far as I could see, and in the shadow of the tower, the earth seemed to be bubbling with a frothy carpet of feathers.

His colleague, coming off watch, explained: 'It's like being in a coffee percolator full of birds up there [in the lightroom].' It was now Hill's watch so he was able to take a look for himself and as he climbed the tall tower, every window he passed was alive with birds. Even inside the chamber there were at least 30, flapping wildly about, some with broken legs, some with broken wings. They had entered through the vent in the roof which used to let the paraffin vapour escape before Hyskeir went electric. Feathers brushed his hair and cheeks so he protected his face with one arm while he wound the clockwork mechanism with the other. It reminded him of that famous scene from the Hitchcock movie, and indeed it proved a long night's watch.

> The next morning was like the calm after a battle. Most of the birds had gone, but one in a thousand had flown straight into the light, like a moth to fire, and broken their necks on the lighthouse glass ... All of the outhouses had flat roofs, and when we removed the dead ones from around the light, and caught and released those still alive and frightened in the chamber, we looked down from the tower and saw that the flat roofs were carpeted with dead birds. It was one of the saddest sights I'd ever seen …

Seventeen-year-old Tony Marr's experience took place at the other end of the country from Peter Hill's. Being scrupulous recorders, the boys were more conservative in estimating numbers being attracted to the light. As keen birders they were able to able to identify over 20 species, most of them passerines. By contrast Peter Hill on Hyskeir, as a layman, gave a less specific but more graphic account of a similar event – typical of, though perhaps more eloquent than, the average lightkeeper. To most lighthouse men such bird rushes were familiar, if not necessarily frequent. Some keepers were remarkably adept at identification, doubtless others were a bit hesitant about all but the most obvious species; to some, 'wrens' included warblers and goldcrests. In 1884, when a keeper at Tushkar light in Ireland received a copy of Morris's *British Birds*, he confessed that 'all birds heretofore entered as titmice were probably willow or other warblers.' The ornithologists analysing the results had to take records on trust, but it is remarkable how skilled many keepers became.

Hyskeir lighthouse off Canna

To most lighthouse men however, even those with a strong ornithological bent, bird strikes were a nuisance, at best requiring a clean-up in the morning and providing a few tasty morsels at mealtimes, at worst causing damage to the lighthouse apparatus. Hill, and most other keepers, would be unable to identify many of the species involved but like some of his colleagues elsewhere, he was only too conscious of just how many birds had been killed that night on Hyskeir – one of the saddest sights he had witnessed.

An early contemporary of Marr and Hill's, who had a foot in both camps, came to be stationed at the lighthouse on North Ronaldsay in Orkney in 1964. Although employed as a lightkeeper, Ken Walker initiated regular ornithological recording on the island, which has continued ever since. Walker contributed the final chapter to Mary Scott's *Island Saga: the story of North Ronaldsay* (1967), describing the island's birdlife. He recorded finding a dead lesser grey shrike in a water barrel on 30 May 1965, along with a house sparrow (both considerable rarities on the island). And on 8 June 1965 he caught an even rarer Alpine swift.

Walker found that skylarks were seen in large flocks in spring but much less numerous in the autumn. Of all the birds killed at the lighthouse none exceeded redwing in numbers. He also concurred with Robert Clyne in finding starlings coming to the light in large numbers but with relatively few actually being killed. Walker received ringing training from the British Trust for Ornithology and on 16 October 1964 he caught a migrant starling which had been ringed in southwest Norway on 6 September 1963: another he caught and ringed at the lighthouse on 31 October 1964 was recovered back in Norway on 25 April 1965.

According to Kevin Woodbridge, who came to North Ronaldsay as the local GP and helped found the Bird Observatory there, Ken left after a couple of years to work at the seaweed processing factory in Lewis: he died in 2008.

Reflecting upon the tragic irony of bird strikes in Ireland, Richard Barrington wrote how:

> The brilliant lights which save thousands of human lives by warning our vessels of danger, attract, on the other hand, thousands upon thousands of winged voyagers, numbers of whom, when bewildered by the glare on dark nights, fly oftentimes against the lanterns with great force, and are killed striking.

The earliest bird strike incidents I have located so far date from 1879, but without doubt they had been occurring ever since powerful lighthouse beams were invented. Indeed, Henry Wadsworth Longfellow (1807–1882) wrote a poem *The Lighthouse*, published in 1849, and he was obviously familiar with the phenomenon:

> *The sea-bird wheeling round it, with the din*
> *Of wings and winds and solitary cries,*
> *Blinded and maddened by the light within,*
> *Dashes himself against the glare, and dies.*

Sumburgh Head lighthouse, Shetland

In 1879 Mr Anderson, keeper at Sumburgh Head lighthouse described in his very first schedule to Cordeaux and Harvie-Brown how 'in twenty-one years I have not seen so few birds strike the lantern.' This obviously implies that he had been aware of the problem for some time. That year the amenable weather conditions seem to have been the reason. Prior to 1879 lightkeepers had lacked any outlet to record such experiences for posterity – other than perhaps the log books kept in all light stations: and I have not consulted these.

It is easier to find early references to bird kills in North America on the internet. The 34 m (112 ft) tall Cape Hatteras Lighthouse, North Carolina, was built in 1802. Ducks, geese and other species were soon noted as flying into the windows, cracking the glass. A poetess Celia Thaxton lived on White Island in the Gulf of Maine as a ten-year-old, from 1845 to 1847, and wrote:

Many a May morning have I wandered about the rock at the foot of the [lighthouse], mourning over a little apron brim-full of sparrow, swallows, thrushes, robins … and many more besides – enough to break the heart of a small child to think of. Once a great eagle flew against the lantern and shivered the glass.

A survey of some 60 lighthouses in the USA in 1878–9, had 24 replies of which 20 indicated birds struck the lantern sometimes only seldom. Cape Hatteras, Cape Cod, Massachusetts and Cape May, New Jersey, were amongst those at the other extreme. Reports from America, Canada, Australia, New Zealand, Denmark and elsewhere are to be found subsequent to this.

Once ornithologists in Britain came to realise what was happening at lighthouses every spring and autumn, they mainly relied upon lightkeepers to submit specimens for examination, to add to their species lists. The migration schedules they were then asked to complete, together with records published ornithological journals at last provided suitable outlets for reporting while, in turn, many keepers were becoming more adept at identifying the various species they were seeing. Furthermore, after 1900 or so we see a few keepers committing their life experiences and reminiscences to paper.

In October 1901, for instance, John Campbell on Bell Rock recorded how redwings, fieldfares, blackbirds, larks, starlings, wheatears, finches, tits, woodcocks and even the tiny gold-crest, could be found dashing themselves against the glass (see Chapter 6). Campbell's successor on Bell Rock, Robert Clyne, also came to know how the lighthouse light attracted many a bird (Chapter 10). One of the more dramatic and distressing descriptions I have found was from Skerryvore, when James Tomison noted the first great rush of the season on the night of 20/21 September 1903:

> At about 8 pm on the 20th a few meadow pipits and wheatears were noticed on the lantern, and at 10 pm the meadow pipits were flying round in thousands. Standing on the balcony watching them, one could almost imagine there was a heavy fall of snow, the flakes abnormally big. Thousands were flitting about; hundreds were striking against the dome and windows; hundreds were sitting dazed and stupid on the trimming paths; and scores falling to the rock below, some instantaneously killed, others seriously injured, falling helplessly into the sea. This continued till dawn, when all that were still uninjured disappeared ...
>
> On 21 October [the next night] there was a great rush of redwings, fieldfares, blackbirds, and thrushes, with a few starlings ... From 7 pm till dawn the following morning their numbers were far in excess of anything seen here for years. As the wind was strong, nearly every one as it struck was carried away, falling into the sea, a small percentage only falling and remaining on the balcony. Yet in the morning we picked up 98 dead on the gallery. Watching them from the lee side of the lantern, from 10 pm to midnight there seemed a constant fall of dead and maimed ... Sometimes we use the terms hundreds and thousands without thinking what these figures mean, but on this occasion when I say thousands were killed I do not exaggerate in the slightest. Unfortunately, that night there was a very heavy sea washing right over the rock, so that not a single specimen was left around the base of the tower ... The rush consisted mainly of fieldfares and redwings, with a few thrushes, blackbirds, and starlings.

Robert Wilson, another of Eagle Clarke's contributors to *Scottish Naturalist,* wrote of an enormous rush of redwings while he was keeper on Skerryvore. The fall took place almost every night from 17 to 25 October 1914.

> What interested me most was the terrible number of birds killed by striking the lantern. The largest number of dead for one night was 131, counted on the morning of 22nd October; these were gathered on the balcony, the narrow strip of grating surrounding the lantern and the gutter round the base of the dome. Undoubtedly,

the enormous majority of those killed fell into the sea. We saw dead birds drifting away to leeward, in a steady stream, on which many seagulls fed. We watched this, morning after morning; it was really awful …

From the 17th to the morning of the 23rd this slaughter continued, and on the night of the 23rd the birds arrived as usual, but on the following morning we only found one dead redwing on the balcony. The weather was the same as on the former night, viz. south-east wind, cloudy sky with light haze, the only difference being that the wind was light, instead of fresh. This inclines me to think that the strength of the wind may have something to do with the heaviness of the death-roll. Possibly when a strong wind is blowing, birds coming under the influence of the light only become aware of their danger when it is too late for them to arrest their rapid flight, and in consequence many more are killed than when the wind is light.

At first, quite understandably considering the vital importance of their duties, lightkeepers were more concerned at the damage birds inflicted on their apparatus. On the 6 October 1902 John Campbell on Bell Rock noted:

… our first intimation of the autumnal migratory flight in the arrival of a flock of wheatears, accompanied by a solitary wren. On the 27th several greenfinches, larks, and starlings were making insane efforts to follow the line of *most* resistance, resulting in our new lantern receiving its first baptism of blood, as the glass next morning testified.

Later he added:

The found killed amounts to over 200, few in proportion to the estimated number visiting. Grease, lime, blood and feathers half obscure the lantern panes, and all require liberal cleaning to keep up to inspection order.

But in October 1901 he had described more serious damage:

… the heavier birds do not always strike with impunity; instances have occurred where ducks have gone clean through the lantern to the derangement of the revolving gear of the light, the splintered glass bringing the machinery to a dead stop. An incident of this nature happened a few years ago at Turnberry Lighthouse, on the Ayrshire coast, the intruder in this case being a curlew or whaup. A storm-pane is considered a necessary adjunct to every lightroom, and is always held in readiness to be shipped in case of such emergency.

As long ago as 1827 a flock of wild ducks hit the glass panes at one of the Lundy lighthouses in the Bristol Channel. This necessitated emergency repairs before replacement windows could be sent out to the island; seven dead were picked up on the gallery outside; doubtless they were not wasted!

Robert Clyne on Bell Rock was cautious when opening the balcony door to prevent starlings or thrushes flying in and damaging the burners (Chapter 10). Peter Hill witnessed birds entering the lightroom down the vent that let out the paraffin fumes. If a door was left open, Clyne reported, birds such as curlew just walked in but made no attempt to fly once inside. He also recorded a jack snipe with a broken leg that had come in an open door and a willow warbler that had come through the revolving cowl of the lantern vent and rested aloft all night until the light was turned off next morning. A swallow flew in an open window of the kitchen one night and sat on one keeper's head as he was having breakfast! These uninvited visitors did not cause a serious problem to the delicate machinery but Clyne did resent a storm petrel that vomited all over the glass outside, the smell persisting for some time despite numerous washings and rain storms.

Campbell recorded how at some shore stations it was customary on the approach of a likely night during the migratory period, to keep the cats indoors to prevent them mangling the expected catch. Several keepers noticed birds of prey – such as short-eared owl, tawny owl, kestrel, peregrine – flying around outside during a rush of birds at a lighthouse, picking off victims. If the migrants survived trial by spotlight they risked falling prey to a hungry predator.

One night in 1881 H Knott, keeper of the Skerries lighthouse off Anglesey, found the lantern gallery full of birds and more casualties next morning at the foot of the tower; it necessitated the use of a wheelbarrow to remove them to the station garden for manure. At Hyskeir Peter Hill mentioned having to clear carcases from the flat roof of some of the buildings next morning. The trumpet-shaped foghorn too often demanded similar treatment; one keeper on Bardsey recounted having to remove some two or three dozen desiccated bodies of grasshopper, sedge and willow warblers.

John Campbell on Bell Rock told how the birds he collected around the light one morning filled an ordinary clothes-basket; a few nights later these included five wild geese, out of a large flock that came to grief on the dome. Doubtless such unexpected bounty was welcome: Campbell noted how fieldfares, blackbirds and redwings, sometimes even a woodcock, could grace the dinner table during the migratory season (Chapter 6). Robert Clyne confirmed how in his young days how '…the most interesting time of bird migration was the arrival of the woodcock in October or November and good sport, and as good cuisine, was to be had for the days they remained'. At Eddystone, Eagle Clark had similarly reported how '… the bodies of the various thrushes and skylarks were served up at dinner for several days, and proved a most welcome relief from the tedium of salt beef …'. At Little Ross in 1917 William Begg was able to pick up dead lapwings, which he made into stews and pies. Back in 1881, some workmen working for Knott at the Skerries Lighthouse had made 'a monster pie' from two hundred skylarks and other smaller birds!

The species involved in falls, rushes and strikes were diverse of course, and the schedules submitted by lightkeepers mention at least 60, mostly land birds. Harvie-Brown clarified:

Left: Starling

Right: Young Lapwing at night

'Next to the lark, the starling occupies the most prominent position in the reports ... ' But without going into the huge detail contained therein to demonstrate the diversity, it may suffice to recall the eloquent writings of two, now familiar, lightkeepers in particular – Robert Clyne and John Campbell. Clyne wrote how:

> ... the majority are starlings, but there are also a good many blackbirds, thrushes and larks. Few starlings were killed in proportion to their numbers [an observation repeated by Ken Walker on North Ronaldsay] ...

Clyne also listed redwing, fieldfare, Norwegian nightingales [song thrushes?], robins, goldcrests, a bluethroat, willow warblers, swift, storm petrel and jack snipe.

> In the spring migration, the first visitors to the lantern are usually the wheatears; then follows the warbler family. Others of the multifarious families I have seen killed or caught on the lantern are the cuckoo, land rails [corncrakes] and water rails, water-hen [moorhen], whinchats, flycatchers, and pipits. Of the finch family not many visit, but I have seen the linnet, siskin, the little redpoll, brambling, and chaffinch. Of seabirds and other residents I have found in the casualty lists are ducks, gulls, sandpipers, lapwing, golden plovers, snipe and curlew.

John Campbell added:

> Amongst the more noteworthy of our captures here, at various times, the following may be mentioned: – a peregrine falcon, large horned owl [long- or short-eared?], small brown owl [tawny], kestrels, sparrowhawks, crows, cormorants, corncrakes and a turtle dove.

In his book *Fair Isle and its Birds* (1965) Ken Williamson gave a vivid description of a rush one misty autumn night at the island's South Light; a comparatively rare event there.

The hazy ink-blue sky shuddered with movement, hundreds of confused birds being held prisoner in the revolving beams … the hundreds of fieldfares looked strangely beautiful as the glare caught the shining white undersides of their wings. It struck me as strange that blackbirds were hardly ever to be seen in the rays, though they were the commonest birds at the lantern; probably their uniformly dark plumage, lacking all contrast, explains their cloak of invisibility. Redwings were common, song thrushes less so, and we took a few skylarks, starlings and bramblings in the light. Occasionally a long-billed, cumbersome woodcock flew past, looking huge in the beams, and all night long a lone common gull flew round and round the tower, ghostly at the end of dim tunnels where successive beams pierced the haze a hundred yards away. It must have flown many, many miles without getting anywhere.

Fair Isle South at night

In his Irish report, Richard Barrington included a list of species prone to striking – not necessarily typical of British lighthouses. The top 12 were as follows:

Blackbird	148	Willow warbler	82
Sedge warbler	142	Starling	80
Skylark	130	Redwing	61
Song thrush	118	Goldcrest	56
Whitethroat	109	Chaffinch	47
Wheatear	89	Fieldfare	43

All 12, he noted, were passerines, a fact that also intrigued Eagle Clarke. All the birds attracted to the Eddystone and Kentish Knock lights in the latter's experience were passerines. But he could not think of any reason why songbirds should suffer such a fatal attraction.

The species list is likely to vary according to locality; for instance southern lighthouses will attract some continental birds that the northern ones may rarely encounter. Not all birds migrate so may never be recorded at a lighthouse, and it is interesting that, amongst those that do migrate are some that just do not seem attracted to lighthouses. Interestingly, in Kircudbrightshire William Begg commented how: 'It is wonderful what an attraction the light has to most land birds, yet there are others which it never affects. The rock pipit, for instance, though locally very numerous, rarely appears at the light.'

Clyne mentioned how in his experience swallows were never attracted to lights and William Begg concurred except that in all his 32 years as a lighthouse keeper, he had never seen swallows at any lantern except at Little Ross.

On an encouraging note, a keeper at Kinnaird Head offered: 'birds are every year getting more scarce, as the town is now extended to the lighthouse.' John Campbell agreed up to a point, noting how numbers of arrivals at Bell Rock decreased each year, but speculating that it may be due to the rise in number of lights on the coast (Chapter 6).

It had long been evident to keepers that some lights proved more attractive to birds than others. From the light characteristics described in the keepers' schedules Harvie-Brown and Cordeaux were able to deduce: ' … the brightest, whitest fixed lights have the most influence in penetrating fog or haze, and therefore of attracting birds.' Robert Clyne noted that if this were so, the light on the Isle of May, being more powerful and with more flashes, should be more fatal than that of Bell Rock (Chapter 10).

At the Casquets in the Channel Islands, which is a revolving light, it was reported how 'the larger birds follow the rays, but do not often strike the glass.' Apparently in 1961 when the revolving beams of Dungeness lighthouse were replaced by a bluish-white occulting light (producing a one second flash every ten seconds) there was an immediate and noticeable decline on its attraction to migrants. The tower itself is now also floodlit at night as an additional aid to shipping and to reduce bird losses.

Left: Skylark

Right: Rock Pipit

Harvie-Brown had noticed how lightships always returned the fullest schedules, and therefore a preponderance of birds was noticed all along the East English coast where lightships were most abundant. The East Scottish coast returns were smaller there being no lightships at that time, but only lighthouses, whose lanterns were at a greater height.

One of Barrington's correspondents from Barrels Rock lightship, Ireland, had led him to comment how at '… this very quick red light, flashing 30 seconds, no birds ever struck the lantern. I believe the quick revolution frightens them away, as in fog and thick weather no birds ever stay about the light like the ships that have the bright light.' Another keeper on Coningbeg lightship added: 'very few birds killed against lantern since the light was changed from steady to flash light.'

But the colour of the light also played an influential role, red seemingly being less attractive. The Bell Rock, although a flashing white light now, used to have a red element in its character. John Campbell recorded once how a large group of gold-crests, circling around the lantern, honed in only on the white flash, and not the red (Chapter 6). Such observations convinced Eagle Clarke that, contrary to popular opinion, the birds that appeared at a lantern are not lost and making their way to the light in the absence of any other directive impulse. They were positively attracted to the light.

At Eddystone, the emigrants which I saw in such numbers had practically only just left the land behind them, and had not had time to get lost when they appeared at the lantern. Another important fact in support of my contention is that the birds never appear at the light-stations at night, except when the rays are remarkable for their

Casquets lighthouse, and the gannets on Ortac, Alderney, Channel Islands

luminosity … Another significant fact is that they do not seek stations having red or green lights. Such lanterns, I am informed by the keepers, are seldom if ever visited under any conditions … When the Galloper lightship [east of the Kentish Knock] had white lights, great numbers of birds were allured to its lanterns, but now that the light is red, bird visitors are almost unknown. If the birds were lost, why should they seek a white light, and avoid one that is red or green? That the migrants may and do become confused, and for a time perhaps, lost after the excitement and fatigue occasioned by their attendance upon the lantern, I can well imagine.

It was obvious to all concerned that more birds were attracted to lights – and therefore more killed – in autumn than spring. Barrington thought on longer autumnal nights as a factor but also ventured – probably rightly – how after the breeding season there are more

birds involved in migration, and that a high proportion of them are young, inexperienced birds more likely to succumb to hazards. It is well known that hosts of birds die on migration, at lighthouses certainly, but many, many more die from other causes such as exhaustion, starvation and predation for example, especially if they encounter bad weather on route. Some may be drowned at sea or blown off course, never to reach their destination.

That the weather was an important factor in lighthouse incidents was well known to keepers. Robert Clyne mentioned how 'there may be several rushes of migrants during the autumn dependant mainly upon the weather.' At Skerryvore James Tomison observed how:

It is pretty well known now by all students of bird migration that the beams of light issuing from the lantern of a lighthouse attract the passing flocks, and this especially is the case when the atmosphere is hazy, and when there is a night of 'sma' rain or Scotch mist. The condition of weather undoubtedly makes the beams more conspicuous and attractive.

Clyne continued:

The night must also be quite dark. Moonlight is a most unfortunate time for the observer who is anxious to witness a rush, for I never yet saw a bird on the windows when the moon reigned on high. Also if the atmosphere is very clear, no matter how dark it may be, the passing crowds pass on without a pause. But in spring or autumn, if the wind is easterly, and the horizon is hazy, hazy enough to obscure a light about eight or ten miles distant, and if a large crowd of birds happens to be passing, the scene witnessed from the balcony of the tower is really worthy of being termed one of Nature's wonderful sights … In ordinary clear weather they pass at a great height, so high as to be invisible to the naked eye.

In October 1901 John Campbell on Bell Rock confirmed how on a hazy moonless night, a southeasterly breeze and drizzling rain brought the autumnal migratory flight.

To take Tomison's first point about moonlight, Barrington summarised from the Irish schedules. Of 673 thrush specimens received, only 106 (less than 16%) were killed when the moon was more than half full. Furthermore, 32 snipe, jack snipe and woodcock struck on the first and fourth quarter of the moon, but only four in the second and third quarters; as with many other species 90% struck within a week of the new moon. Harvie-Brown and Cordeaux went on to conclude:

When the nights are dark and cloudy, no stars appearing, in rain, fogs and snowstorms, flocks of birds during the night migrations will crowd round the lanterns of the lightships; many strike the glass and are killed, falling on deck or pitching overboard. On these nights birds will often remain for hours in the vicinity of a light, circling round and round, evidently having lost their way; at the first break in the clouds, the stars becoming visible, or the first streak of early dawn, they will resume their flight to the nearest land.

From Skerryvore, Robert Wilson witnessed how the wind influenced mortality at a lantern:

… the strength of the wind may have something to do with the heaviness of the death-roll. Possibly when a strong wind is blowing, birds coming under the influence of the light only become aware of their danger when it is too late for them to arrest their rapid flight, and in consequence many more are killed than when the wind is light.

In moving from station to station in the course of their career, keepers – as we have seen – were aware that not all lights caused a problem, some being worse than others. While lighthouse keepers' comments were all distinctly relevant to the phenomenon, relatively few scientific assessments of bird strikes appear to have been undertaken. One exception is at the lighthouse on Bardsey, 3 km (c 2 miles) off the end of the Llyn Peninsula in North Wales, recognised as having one of the worst casualty rates in Britain. The issue has been recorded in some detail ever since its Bird Observatory came into existence in 1953.

One of the most remarkable falls of migrants witnessed at the Bardsey Lighthouse was on the 9/10th November 2002, as one of the Observatory's ornithologists, Steve Stansfield, recounted in *Bird Observatories of Britain and Ireland* (2010).

The moon was five days old (28.6 % waxing). There was full cloud cover with heavy and persistent rain and visibility was poor. Twilight had ended at 1853 hrs, though the moon rose at 02:06 hrs. Birds were first seen in the light at 0030 hrs and the attraction was attended from 0100 hrs. Very large numbers of birds, with many fatalities, had already occurred when observers arrived … the large number of birds dropping into the illuminated area was reminiscent of watching a blizzard. A decision was taken to collect only chilled and injured casualties due to the number of birds which were grounded … The area in front of the gantry resembled a bucket of maggots – shimmering and moving. The noise was tremendous. Numbers were estimated at 30,000 redwings, 500 fieldfares, 350 blackbirds, 300 song thrushes, 350 starlings, 30 lapwings, nine golden plovers, 10 woodcocks, seven dunlin, seven snipe, four redshanks, two blackcaps, two water rails, a storm petrel and a jack snipe. In total 31,573 birds of 18 species were recorded with, surprisingly, only 249 fatalities.

Bardsey is known as Ynys Enlli in Welsh – island of tides or eddies. Opposing currents combine with wind effects to make its surrounding seas hazardous and the island is also encircled by jagged rocks. So its lighthouse came to be constructed to guide shipping safely through St George's Channel and the Irish Sea. Its 30 m (99 ft) square tower, now painted red and white, was constructed by Trinity House in 1821 at the flat, southern end of the island. Its white paraffin beam produced five flashes every 15 seconds. In March 1966 it was converted to electricity, increasing its power to the present 90,000 candelas, visible from 48 km (30 miles) away on clear days. But of course, on clear nights birds are rarely attracted anyway.

In fact the conversion of Bardsey's light to a more powerful beam did not seem to increase the annual death toll. This seems at odds with a study at Long Point lighthouse in Ontario, Canada, when it was automated in 1989: making the light beam narrower and less powerful brought about 'a drastic reduction in avian mortality' from 200 fatalities each spring to 19, and 400 to 10 in autumn (Jones and Francis 2003 in the *Journal of Avian Biology* 34: 328–333).

As we have seen the worst culprits are fixed white lights and those whose flashes are produced by a rotating prism, giving the impression that they are permanently illuminated throughout the night. Occulting lights produce flashes by blocking the light at intervals with an opaque panel and do not seem to be quite as attractive. Interestingly at a lighthouse on the Farallones off the Californian coast hoary bats were attracted on about seven dark, moonless nights each autumn. As soon as the powerful, rotating beam was switched to a weaker, pulsating light, the number was reduced to almost nothing. It seems to have had the same effect on migrant birds although this had not been explored.

When dealing with actual numbers of birds attracted, it is difficult to be accurate. The death toll was reckoned by the Bardsey Observatory to be around 300–350 per annum, but may vary from only 69 to nearly 2,000. Nearly 20,000 fatalities were recorded between the years 1953 and 2008 demonstrating the gravity of the problem at Bardsey. Dungeness in Kent was another hotspot until the light was changed, while lighthouses all round Britain will add to the overall mortality figures, albeit on a lesser scale. Between the years 1886 and 1939 it was calculated that some 500-8,000 birds were killed each year at lighthouses around Denmark – a total of 20,700 in spring and 33,800 in autumn. But relative to the numbers of birds that succumb every season during their migrations (perhaps millions), lighthouse figures on their own – undoubtedly underestimated – are still unlikely to be hugely significant.

But then, around the world, birds die at many other man-made structures every spring and autumn; one study estimated nearly half a million, out of a 10 billion total. For instance, great losses have occurred at tall, illuminated masts used for radio, television and mobile phone transmissions, especially in North America. There, in the 1970s, an estimate was put at 1.3 million per year but by the year 2000 this was increased to 4 or 5 million due to a four-fold increase of such towers. Ceilometers – searchlights designed for measuring cloud height at airports – attract birds, which then collide with nearby buildings. At one such facility in Georgia an appalling 50,000 birds of 53 species were killed in one night. As with

Left: Grey Heron

Right: Manx shearwater

lighthouses, mass mortalities are most frequent on nights of low cloud and fog or rain, when birds fly lower than normal.

Gas flares on oilrigs also attract birds on dark foggy nights; up to several thousand per night in some cases, while another more conservative estimate for the North Sea put the figure at a few hundred per rig per annum. Modern wind turbines and wind farms are recent additions to the collision risk whose effects have yet to be assessed fully: one early estimate in the USA gave a figure of 33,000 annually. At the end of the day, any postulated figures for deaths at man-made structures are surely underestimated since injured birds may die away from the point of collision, be carried away by nocturnal scavengers or washed out to sea never to be recorded.

The Bardsey list of casualties covers 89 species, ranging from birds as large as grey heron to those as small as yellow-browed warbler and from species as common as chaffinch to those as rare as grey-cheeked thrush. The variety is perhaps unsurprising, given that so many species on the British and Irish lists engage in migrational and seasonal movements – local or distant. In common with some other published lists – though not Barrington's above – redwings top the list by a considerable margin, as typified by Bardsey figures.

Bird species most frequently killed at lighthouse attractions at Bardsey from 1979 to 2005.

	1979–1996	*1997–2005*	*Total 1979–2005*
Redwing	3,640	421	4,061
Willow warbler	698	169	867
Manx shearwater	379	241	620
Sedge warbler	338	157	495
Starling	441	23	464
Blackbird	203	20	223
Grasshopper warbler	145	52	197
Song thrush	126	31	157
Fieldfare	97	13	110
Whitethroat	72	35	107

All the corpses collected from the Bardsey lighthouse are stored in a deep freeze and later sent to the National Museum of Wales in Cardiff. Some of these have been put to use in displays at the museum, whereas others have been used in studies of fat levels and body condition. Fat is the fuel utilised by birds during their migrations, and it comprised 5–15% of the weights of 13 sedge warblers killed on the night of 6/7 September 1969, 8–20% for whitethroats and 10–15% for 14 grasshopper warblers. There is a physical limit to how much fat reserves a flying bird can carry on its migration and most will need to stop periodically, en route for Africa, to feed and refuel. Clearly any pointless use of precious fat reserves fluttering aimlessly around a light beam all night severely compromises a migrant's chances of successfully completing its journey – an will ultimately add to the overall mortality figures.

Typically the largest attractions of Manx shearwaters on Bardsey involve adult birds around the new moons of June, July and August. The species breeds on the island – maybe a couple of thousand pairs – but only occasionally is a locally-ringed shearwater caught at the lighthouse. Thus the birds coming to the light were not local breeders, which presumably might be used to the dangers. I strongly suspect most would be wandering pre-breeders. While these tend to return to their natal colony to nest, those attracted to the light would certainly involve birds from other colonies, which is known from ringing to be a common phenomenon. Small breeding colonies exist close to hand on Copeland and Rathlin in County Antrim, while some are now back nesting on the Calf of Man from where they derived their name. Major colonies are also found in Skokholm and Skomer in South Wales not so very far to the south. I used to live on the Isle of Rum in the Hebrides, where there is a very large – and unique – mountain-top colony. On wet and misty nights numbers of adult shearwaters used to get attracted to house lights in the village, or to the lights of boats moored out in the bay, and then in August and September when the young were fledging. More would end up at the harbour lights in Mallaig, 24 km (15 miles) away. Fledging puffins have been similarly attracted to the bright lights of the radar base on St Kilda.

The mechanism of attraction still remains unclear, but presumably the night-migrating birds are attracted to a focal point of light when disorientated – this can happen when their navigation aids (the stars) are obscured by clouds at a time when the birds are in full flight. Once attracted, the birds seem confused and unable to cope with this unusual situation except by flying round the light source, and this they continue to do – be it at a gas flare from an oil production platform or in the powerful beam of a lighthouse – until (i) they find a safe spot to land and rest; (ii) they are killed or injured by flying into some solid object; or (iii) the cloud disperses to allow them to see the stars and resume their migration.

Mention has already been made of the role of the moon. Typically – as lightkeepers, Harvie-Brown, Cordeaux, Barrington and Eagle Clarke all recognised – birds are seldom attracted during full moon periods. We now know that, even with full cloud cover, heavy rain or fog, birds are still able to navigate, using the visible navigational cue of the moon. As a result, there are two weeks in each month when the moon is visible and it is fairly safe for lighthouse observers, such as those on Bardsey, to go to bed and sleep, as – it can be fairly confidently predicted – no attraction of consequence will take place. During the new moon period (usually up to one week either side of the new moon) however, things are quite different. When the weather is conducive, some species begin their migration at dusk. If the weather ahead of the birds deteriorates, obscuring their navigational cures, they may be attracted to a lighthouse. It is significant that most attractions occur on occasions when the waxing moon sets below the horizon during the early part of the night, but has been visible since the fall of darkness (nautical darkness being when the moon is 12 degrees below the horizon and the sun's light is no longer visible). Birds begin to migrate when they can see the moon (even if small) and it provides enough light to illuminate geographical features.

Bell Rock Lighthouse – with (Gail Churchill) and without bird netting

As it sets below the horizon, they no longer have their guiding light and the larger and more spectacular attractions take place.

As we have read, Eagle Clarke had taken netting with him to Eddystone to prevent his specimens being damaged against the glass, but he quickly realised that it actually helped to reduce the mortality by saving many birds which would otherwise have dropped off the balcony and drowned. Some lighthouses have adapted this technique to keep birds from striking the glass of the lantern, Bell Rock being one notable example. When I took a trip out to the Rock in 2009, the netting had been temporarily removed to spruce up the paintwork prior to the visit of the Patron of the NLB, Princess Anne.

Modern studies, such as those conducted by Bardsey Bird Observatory, have not only highlighted the mortality amongst migrants attracted to lighthouses, but have also led to innovative measures to reduce the death toll. Lightkeepers were aware how the railings around the light tower were probably reducing fatalities. As early as the 1940s Trinity House attempted to reduce the number of birds killed at Bardsey and elsewhere (such as South Bishop) during spring and autumn migration by erecting a series of ladder-like perches extending out from the balcony at the level of the light. Unfortunately more birds struck the extensions (and were fatally injured) than actually perched on them.

In the late 1950s and early 1960s, Bardsey tower itself was illuminated by paraffin floodlights to make it more visible during bird attractions, but it is uncertain whether or

South Bishop lighthouse, St David's Head, Wales

not this may have resulted in more birds being attracted to the white sections of the tower. These floodlights were covered by red filters in 1963 in order to illuminate the tower without producing glaring white patches of wall. As early as 1959 an alternative suggestion was made (by Ken Williamson and Reg Arthur) to illuminate an area of grass and gorse a short distance from the base of the tower in order to provide a floodlit area which birds could see and where they could land to rest.

So in 1978, a gantry with two 500-watt floodlights was erected at Bardsey light, the equipment being provided by Trinity House and the RSPB. At first it was very successful in drawing down to the grass and gorse an estimated 4,000 birds, mainly thrushes and starlings, and holding them there until dawn. The apparent success of this system continued in 1979, when the light output was doubled by the inclusion of two more floodlights on the gantry; October attractions saw up to 3,000 birds, again mainly thrushes and starlings, safely on the ground at dawn. Later, in 1979, a gantry with two 500-watt lamps was erected to the east of the lighthouse. The floodlights illuminated the ground on that side, affording the birds a visible area on which to land. Since the inception of this system, it has become apparent that the lights do cause warblers to come down into the shelter of the gorse, and thrushes and starlings onto the grass, thus presumably reducing their wasting of energy through continuous circling in the beams, and perhaps reducing the mortality in these species; with waders, the floodlights have no appreciable effect. During 'lighthouse nights' birds are also caught by hand, and roosted quietly in dark boxes until dawn, when they can be ringed and released.

This single gantry was operated until 1998, when the use of lamps, mounted on three small, mobile gantries, was tried with very positive results. The intention was to bring the birds down closer to the lighthouse tower, and then to move the gantries to the side on which they appeared to be collecting. If, for example, the wind on a given night was from the south, the majority of the birds in the air would tend to be at the northern side of the tower. The gantries would therefore be moved to illuminate the ground on that side, immediately outside the lighthouse compound and 10 m (35 ft) from the tower. By the end of the season, there had been several large attractions, but fewer than usual casualties. At one of these, about 8,000 birds had been coaxed down to safety in the illuminated areas. This system worked to good effect. In 2014 however Bardsey lighthouse was modernised. The optic has been removed to the Henfaes Centre, Porth y swnt National Trust Museum, to be replaced by two red LED lanterns (main and standby) with a range of 18 nautical miles (3o km). Exhibiting a red flash of 1 second every 10 seconds, the unit of powered by a solar system, replacing diesel generators running day and night. This is in accordance with its navigational requirements but has the additional benefit of reducing avian mortality significantly (Trinity House).

The Butt of Lewis was another lighthouse that suffered from occasional bird strikes at its 750,000 candle-power beam. Each morning after an incident, the three keepers had to dangle 81 m (265 ft) above the sea in any weather to clean the panes of glass. Donald Michael's first posting had been on the Flannan Isles shortly before it went automatic in 1971. His last was as principal lightkeeper at the Butt of Lewis before its automation in 1998; he spent four years there and quickly had a brainwave. He acquired two life-sized plastic eagle owls, which he filled with concrete before fixing with steel bolts to the windswept balcony railing. The owls succeeded for a short time until the troublesome migrants lost interest in the motionless scarers and ended up perching on them. Donald Michael retired to Girvan. According to Tony Marr, who nowadays lives close to the Butt of Lewis – and is still birdwatching – its white light flashing every five seconds no longer seems to attract many birds.

Bird observatories such as Bardsey have played a significant role in quantifying and reducing interactions at lighthouses, which, after all, have significant impacts on their very function, as well as carrying tragic consequences for the birds themselves. Hence the willingness of lighthouse authorities to co-operate.

With the contribution made to ornithology by lighthouses it is not surprising that bird observatories tend to be attracted to them, much as migrants are! The concept behind them is as a field station to monitor migrations and bird populations; and for catching, examining and ringing birds to help forward these aims. Bardsey is not unique; indeed there are currently 19 bird observatories established in Britain at the moment, the latest addition also happening to be the very first established in Britain and one of the first in the world – Skokholm. The island lies 4 km (3 miles) off the southwest Pembrokeshire coast; barely a square mile in extent with its Old Red Sandstone cliffs rising in places to 50 m (164 ft), 5 m short of its highest point. A lease was taken up in 1927 by farmer/naturalist Ronald Lockley (1903–2000), who established the Bird Observatory six years later. He had been impressed by a visit to Heligoland organised by WB Alexander (1885–1965), who also helped inspire the British Trust for Ornithology and its ringing scheme.

Skokholm lighthouse, southwest Wales

A simple Heligoland trap on Fair Isle

Lockley saw the application to Skokholm, and he began constructing the first bird trap in his cottage garden. It was the same design as developed on (and named after) Heligoland by ornithologist Max Hugo Weigold (1886–1973) when he worked at the Marine Research Station on the island, and later became Director of Hanover Museum. Lockley's Heligoland trap was a more modest sized funnel of wire mesh, into which birds could be coaxed to a small glass-backed wooden catching box at the end.

Skokholm Bird Observatory, at first Lockley's own private initiative, had to be abandoned during the Second World War before ringing ceased in 1976. By this time the lease had been assumed by West Wales Naturalists' Trust who, in 2006, succeeded in purchasing the island. Two years later it was declared a National Nature Reserve, but in the meantime many important research projects had progressed, not least a long-term study of Manx shearwaters first initiated by Lockley himself. In 1976 I was privileged to show Ronald Lockley the shearwater colony on the Isle of Rum.

Skokholm Bird Observatory was based in the cottage farmhouse close to the lighthouse, built in 1916 and automated in 1983. During its conversion to solar power the light was changed from red to white, which immediately led to bird strikes. However red shades were introduced to cover the island sector of the beam, which seems to have overcome the problem.

In 1934, following an advisory visit by Alexander and Lockley, the Isle of May became Britain's second observatory, based in the redundant Low Light. Skokholm and the May were to remain Britain's only two bird observatories until after the Second World War.

Although the Isle of May was Scotland's oldest lighthouse (built in 1636), its coal-fired beacon was not replaced until the Commissioners of Northern Lighthouses bought the island in 1814. Two years later Robert Stevenson built his grandiose new light. The fixed light lantern was replaced by a revolving one in 1844 when the Low Light was built. With the deployment of the North Carr lightship 11 km (7 miles) to the north at the mouth of the Firth of Forth, the Low Light became surplus to requirements. In 1886 electricity was installed to replace the old oil light on the main tower – the first to be powered by electricity in Scotland. As Joseph Agnew's prolific schedules had revealed (see Chapters 9 and 10) the Isle of May had gained a reputation for attracting and killing large numbers of birds until 1924 when the expensive electric light converted to paraffin. Later reduction in candle power further minimised bird casualties. The lighthouse families moved to the mainland in 1972 following redesignation as a rock station.

One of the first naturalists to visit the Isle of May was Charles Darwin in 1826 when he was a medical student at Edinburgh University. But its real potential for the study of bird migration was not recognised until two remarkable young Fife ladies of private means began to visit some 80 years later. Leonora J Rintoul (1878–1953) and Evelyn V Baxter (1879–1959) were close friends – both socially and geographically in that throughout their lives they lived only a few hundred metres apart from one another in Largo, Fife. As teenagers they were birdwatching at Tentsmuir one day when they first met Eagle Clarke, engaged in promoting the place as a nature reserve. Inspired by subsequent visits to his bird collections in the Royal Scottish Museum and by his work on Fair Isle, they initiated

Isle of May Low Light, now the Bird Observatory

comparative migration studies on the Isle of May, 8 km (5 miles) off the Fife coast. Their regular spring and autumn visits were to continue for 26 years, interrupted only by the First World War. Encouraged by Harvie-Brown before he died in 1916, they published their preliminary results from the Isle of May the year the war ended and highlighted their views, contrary to those of Eagle Clarke, on drift migration. As was the custom of their day, they carried a gun and contributed no fewer than 1,200 skins to the Museum collection; many of them were however picked up dead at the lighthouse by the keepers, including the first British pied wheatear. The 'good ladies' frequently published in *The Scottish Naturalist* and later sat on its editorial board with Eagle Clarke. They compiled *A Vertebrate Fauna of the*

Left: Isle of May Bird Observatory bird watchers in April 1977

Right: Leonora J Rintoul and Evelyn V Baxter

Forth (1935) - a series of landmark volumes initiated by Harvie-Brown - but the culmination of their lives' work was their monumental *The Birds of Scotland, their History, Distribution and Migration: volumes 1 and 2* published in 1953, the year Leonora died.

Just as 'the good ladies' of Scottish ornithology had been inspired by Eagle Clarke, so they in turn inspired a young Edinburgh schoolboy, George Waterston (1911-80) from a long-established family of printers and stationers. With some young companions he first visited the Isle of May in September 1932. They had all read articles on Heligoland by WB Alexander, and on Lockley's bird trap on Skokholm. Both eminent men offered to come north in 1934 and, with the sanction of the NLB, helped George and his friends establish the Isle of May Bird Observatory. Two Heligoland traps were constructed, but at the start of the Second World War the Admiralty took over the island and activities had to cease. But on its re-opening in 1946 the NLB permitted the Observatory to move from the old coastguard house to the more commodious Low Light, where it has been based ever since.

After the publication of his *Studies of Bird Migration* in 1912 Eagle Clarke continued to return to Fair Isle until 1921. On Eagle's final visit, he was accompanied by Rear Admiral Stenhouse – who would himself return until he too found the passing years too strenuous for island visits. Stenhouse had enlisted the invaluable assistance of a local crofter called George Stout of Field (1886–1966) (known as 'Fieldy' and not to be confused with Eagle Clarke's pupil, George Stout of Busta). Stenhouse then nominated George Waterston as his successor. George had first visited the island in 1935 and returned annually until the war started, where he stayed at Field. George later recalled how 'at the end of the day it was an education to watch [Fieldy and fellow islander Jerome Wilson] both skinning and preparing specimens, but a bit disconcerting to find that the skinning knife used by Fieldy (with arsenical soap as preservative) was also used for cutting bread.'

George was captured in Crete in 1941 and spent 30 months as a prisoner of war in Germany. Although suffering ill health he made good use of his time with other ornithologically-minded prisoners (some later to become prominent in bird circles after the war), making studies of birds around the camp when they could. By this time he had Ian Pitman, an Edinburgh lawyer and fellow prisoner, fired with enthusiasm for a bird observatory on Fair

Far left: George Stout of Field, Fair Isle

Right: George Waterston

Observatory bird watchers seawatching at the South Light

Isle. In October 1943 – with severe kidney problems – the Red Cross arranged for George to be repatriated home through Sweden. Escorted north through Norwegian waters, then west across the North Sea, the ship's first sight of land - much to George's delight, with a promise of things to come - was Sheep Rock on Fair Isle!

After the war, with Ian Pitman, George was able to buy Fair Isle and enlisted other like-minded friends to set up his long-hoped for Fair Isle Bird Observatory in 1948. Six years later the island was passed to the National Trust for Scotland. The observatory was located in former naval huts, midway between both lighthouses. First arriving in 1966, I used to enjoy stopping at North Light, where I was always welcomed by brother and sister Tommy and Katy Russell - originally from Kiltarlity near my home town of Inverness.

Birdwatchers from the observatory (myself included) liked seawatching from the wall of the South Light. Once a group were intently peering through binoculars and telescopes, when Assistant Keeper Angus Edwards mischievously fired up the foghorn. He felt that if their only idea of fun was to spend hours peering at some poor bird, they obviously needed some excitement in their lives!

Fair Isle Bird Observatory is now in modern accommodation but has continued to function as a private trust, with George as its first Secretary until his early death from kidney failure in 1980. In the post of Scottish Director for the RSPB, George was behind countless other initiatives, such as Strathspey nest protection of the first returning ospreys, and an early attempt to reintroduce white-tailed sea eagles to Scotland. As a schoolboy I met him first when I volunteered at the osprey camp in Boat of Garten, and subsequently a few times on Fair Isle. When in 1975 I became involved in the next (successful) sea eagle project on the Isle of Rum, I kept him up to date with progress until he died.

Now Scotland had two bird observatories, Fair Isle, and the Isle of May – the latter based in an old lighthouse. Following on from Skokholm in Pembrokeshire, the fourth British observatory – Spurn - was founded in 1945, the same year as Fair Isle, and others were to follow. Not all have survived, but 19 still function today. Many of them share locations with lighthouses, and some are even accommodated in lighthouse buildings. Copeland Bird Observatory, for instance, is accommodated in the restored keeper's house built with the lighthouse in 1815. Bardsey was set up in 1953 with the assistance of one of its lightkeepers, Alan Till. Previously he had been a keeper on Skokholm and worked there with Lockley on bird trapping and ringing. From the outset he was an obvious choice to

Fair Isle's first bird observatory with Sheep Rock behind

join the Bardsey Bird Observatory Committee. Although the observatory was situated in a large house in the centre of the island, with traps in its gardens, the assistant wardens were accommodated in old lighthouse keepers' cottages a mile away – which for a time had the only available toilet!

When Portland Observatory was established in 1955, the old Lower lighthouse was purchased as its base. Connected to the Dorset mainland only by the long shingle spit of Chesil Beach, Portland is effectively an island. In 1906 both Lower and Higher Lights were replaced by a new single, 41 m (134 ft) high lighthouse at the very extremity of the Bill, which is still in use today. When it was automated, the former keepers' accommodation became a visitor centre.

thirteen

Still Friendly and Welcoming

*I have all the modern navigational equipment on board, but there is nothing more
friendly and welcoming than the sight of a flashing lighthouse.*

Ocean-going ship's master

War was nothing new to the lighthouse service, but the 20th century conflicts were in
a different league. Fortunately there were no lighthouse casualties within the NLB
in the First World War. Sir Edward Grey, the Foreign Secretary – and keen ornithologist –
had observed at the outbreak of the war that 'the lamps [were] going out all over Europe'.
Although he had not been referring to lighthouses specifically, so it proved. All but the most
vital beacons were extinguished all round the British coast so that the enemy mine-layers
could not orientate. The power of other lights was reduced and some lighthouse towers were
even camouflaged or painted black.

During both wars, keepers would be sent signals if Royal Navy movements were
anticipated nearby. In October 1915 the Bell Rock never received the warning and, with no
guiding light visible, the cruiser HMS *Argyll* (10,850 t) struck the west side of the Bell Rock
reef. She had been making for Rosyth to rejoin the fleet after a refit in Plymouth, and one
of her young midshipmen HE Evans later recalled: 'That night was one of the worst I can
remember in long years at sea – storm-force winds, mountainous seas, thick fog and heavy
rain; as unpleasant a combination as any seaman could face.'

At first Principal Keeper John Henderson and his two assistants, Colin McCormack
and Donald Macdonald, did not realise what had happened and when they did discover
the warship stranded on their doorstep they hid at first, thinking they were about to be
shelled by a German battleship. But soon they were making Herculean efforts to get a line
aboard, during which McCormack became entangled and almost lost his life. Meanwhile
three local lifeboats from the mainland were forced back to port in the heavy seas. Finally
two destroyers arrived to take over the rescue of the *Argyll's* entire complement of 655. Six
weeks later the Admiralty installed telegraphic communications between Bell Rock and
Fife Ness.

Copinsay lighthouse, Orkney

After Rubh Re was established in 1912 near the entrance to the naval anchorage in Loch Ewe, the NLB were under pressure to provide lighthouses for other great naval bases such as Rosyth, Cromarty and Scapa Flow. Maughold Head lighthouse on the Isle of Man was constructed in 1914 (automated in 1992). Copinsay lighthouse in Orkney was built in 1915 on top of a 100 m (328 ft) sandstone cliff, but it was not lit officially until after the hostilities, in 1919. It would become the first to trial the use of helicopters in the transfer of lightkeepers.

On its lonely cliff top, Cape Wrath was often shrouded in mist so it was planned to build a new tower nearer sea level with access by lift down a shaft from the top. Blasting and quarrying began in June 1914 but a dispute with the contractor brought the work to an end and the project was abandoned. Lightkeepers were withdrawn altogether from the Bass Rock and when they returned after the war, 'gannets' [or fulmars?] and a 'cormorant' [more than likely a shag] were nesting in the buildings. On the east Caithness coast Clyth Ness lighthouse was built in 1916, during the First World War, and various other temporary lights

Rubh Re lighthouse, Wester Ross

Cape Wrath lighthouse

Left: Lightroom of Duncansby Head lighthouse, Caithness

Right and below: Brough of Birsay lighthouse, Orkney

or foghorns installed along the coast to warn of dangers during wartime. A temporary light on an iron tower was erected in 1915 on the Eshaness peninsula in Shetland to warn of the Ve Skerries 14 km (9 miles) offshore, but it was torn down after the war.

On the cessation of hostilities, the NLB went on to establish Duncansby Head, Caithness, in 1924 (automated in 1997). The traditional round tower, requiring specially made interior fittings and furnishings, was now abandoned in favour of a square one; even the lamp room became square. David A Stevenson also built a neat, unmanned lighthouse on the low tidal Brough of Birsay, north of Marwick Head on the west side of the Orkney mainland. The handsome white tower, in traditional style and atop a 51 m (167 ft) cliff, resembles a small castle; it was modernised in 1983. During routine maintenance on 24 May 2002 the helicopter pilot, Captain Tony Taylor – 'a big, soft-spoken Englishman' – was tragically killed. There is a small plaque on the lighthouse in his memory.

Finally, a proper lighthouse was established on the clifftop at Eshaness in Shetland in 1929, the last manned station designed by the Stevensons, in this case David A. Unusually, only one keeper was deployed in the square tower, but its beacon proved insufficient to keep ships away from Ve Skerries. The lamp was made more powerful in 1974 when Eshaness was automated but the obvious solution was to build a beacon on the Ve Skerries themselves, which was achieved five years later.

Eshaness lighthouse, Shetland

Notwithstanding black-outs similar to the Great War, the Second World War brought a new hazard to lighthouses and their keepers – attacks by enemy fighters and bombers. Although they tried to remain neutral by not reporting enemy ship movements, many lighthouses were machine-gunned or bombed. There were about 30 Nazi air attacks on Scottish lighthouses but fortunately damage was largely superficial and soon repaired. A few weeks after the war began, Barns Ness was machine-gunned, and on the eve of the

*Barns Ness lighthouse,
Berwickshire*

invasion of Normandy the same happened at Duncansby. (Incidentally, as preparation for the D-Day landings, Trinity House was responsible for marking the Swept Channel routes for the invasion of Normandy, laying 73 lighted buoys and mooring two fully-manned light vessels to indicate a safe route to the landing beaches.)

Bombs were dropped near Skerryvore in mid-July 1940, shattering a few panes of glass and the mantle of the lamp. At the end of October the Bell Rock was strafed and damaged, requiring a temporary light to be rigged. Throughout 1941 attacks continued – at Kinnaird Head, Holburn Head, Rattray, Stroma, Out Skerries, Auskerry and the Bell Rock again. Soon after Angus Macaulay – the principal lightkeeper at the Bell – moved to Pentland Skerries he was to suffer another visit from German planes. They strafed the Pentland Skerries lightroom just seconds after the keepers had left it.

Longships lighthouse, off Lands End, was hit by enemy bombs on 1 August 1941 resulting in extensive damage; luckily no keepers were injured. Tragically however three keepers were killed when St Catherine's lighthouse on the Isle of Wight was bombed. As a result, keepers were asked if they wanted defensive armaments, but they declined.

A German mine struck Tuskar Rock in southeast Ireland in December 1941, injuring two assistant keepers, one of whom died the next day in hospital – one of the first lighthouse casualties of the war. At Out Skerries in Shetland, on Sunday 18 January 1942 at 11.45 am, an enemy plane dropped a couple of bombs which fell into the sea: it returned immediately to score a direct hit on the boatman's house. The building was completely demolished and its sole occupant, the boatman's mother, was buried beneath the debris, sustaining injuries from which she died a few days later in Lerwick hospital.

But Fair Isle suffered the most, with both lighthouses being hit by gunfire and bombs. In the spring of 1941 the North Light was attacked twice and outhouses were demolished by two bombs; fortunately no one was hurt. But in December that year, Mrs Catherine Sutherland (aged 22), the wife of an assistant keeper, was killed and her infant daughter slightly hurt when a German bomber fired at the South Light. A few weeks later on 21 January 1942 the Luftwaffe returned and a bomb hit the main block. The wife of the principal keeper, Mrs Margaret Smith (aged 50) and her 10-year-old daughter Greta were killed. A soldier, William Morris, manning the nearby anti-aircraft gun nearby, was also killed. Some of the bomb craters can still be seen nearby. In a remarkable response to duty, the assistant keeper at Fair Isle North, Roderick MacAulay, walked three miles through snow and gale force winds to help restore the South Light; he was awarded the British Empire Medal.

There were only two further attacks, both in 1942 at Sule Skerry. But the area near Scurdie Ness, just north of Arbroath, had been attracting undue attention from German bombers and the locals accused the bright white lighthouse of attracting them in. So the poor keepers were instructed to paint the whole structure black.

Sometime during the Second World War, a British aircraft crash-landed on Copinsay just below the lighthouse but it was soon dismantled by the RAF and carted away. Similarly at Barra Head – but unknown to the lighthouse keepers at the time – a Blenheim bomber crashed into the cliffs; the wreck was not discovered until many years later by a rock climber. On Fairisle a German Heinkel reconnaissance plane was brought down by two RAF Hurricanes in January 1941. Exposed and vulnerable lighthouse keepers performed many acts of courage and bravery in wartime. One stormy night in 1944 an American Liberty ship, *William H Welch*, missed the entrance to Loch Ewe and went on the rocks. Two keepers from Rubh Re Lighthouse trekked through slush, snow and peat bogs to assist, and were able to help rescue 14 of the 74 crew.

After the war, even without the Stevensons, lights and beacons continued to be constructed. In 1958, the last manned lighthouse in Scotland was built by NLB Engineer Peter Hyslop at Strathy Point, Sutherland, on the north coast of Scotland. It had first been proposed in 1900 but vetoed by Trinity House and then, on appeal, by the Board of Trade. The lamp is small and compact yet produced a beam 1,600,000 candlepower. The buildings are laid out in a hollow square with covered passageways for maximum shelter from high winds. As with Duncansby, the traditional round tower was abandoned in favour of a square one, 14 m (46 ft) high. It was the first Scottish station built for all-electric operation, with a

Left: Plaque commemorating the casualties at Fair Isle South in the Second World War

Right: Strathy Point lighthouse, Sutherland

major light and fog-signal and was one of the last to be automated in 1997. Just recently, after a detailed review and consultation by the NLB, it was decided that nowadays most shipping in the Pentland Firth passes much further north, while local fishing and leisure traffic is minimal, reducing the need for Strathy Point altogether. Therefore it was permanently discontinued on 2 March 2012 and the station has been sold off – a sad foretaste perhaps of further rationalisation, economics and ultimately closures in the future.

A small prefabricated, manned lighthouse had to be re-established on Calf of Man island in 1962, two years after Chicken Rock went on fire, giving the Isle of Man the greatest density of lighthouses anywhere. Admittedly the two earlier ones were long defunct while Chicken Rock flashed proudly, three-quarters of a mile offshore. Sealed beam units were installed in the new light, adapting the original catoptric reflector system similar to a car's headlamps. They had first been deployed at Barns Ness on the Forth – the first in the British Isles and probably the first anywhere in the world; now they are a major feature of Scottish lighthouses. They usually have three settings – 780,000 candlepower, 960,000 with 1,200,000 switched on in poor visibility. Chicken Rock was automated in 1961/2 and the 'new' Calf of Man lighthouse in 1995. Since then, with the agreement of the Isle of Man authorities, the NLB upgraded Chicken Rock light to a range of 21 nautical miles (34 km) in June 2007, so that the Calf – a possible source of confusion so close by – could be discontinued thereon.

Left: Chicken Rock, Isle of Man

Right: The author with cruise passengers off Chicken Rock

On June 1972 the Government despatched 'a combined expedition' to Rockall, a tiny rock more than 400 km (250 miles) west of the Outer Hebrides. The remit was to 'set up a flashing navigational beacon' as an aid to shipping. There was little doubt that the motive also included a fair degree of politics due to the possibility of oil discoveries in its waters. So immediately it assumed an importance greatly in excess of its size – only 30 m (100 ft)across and 24 m (78 ft) in height. Parliamentarians have commented: 'There can be no place more desolate, despairing and awful … More people have landed on the moon than have landed on Rockall.'

That may no longer be true as in recent decades there has been an unseemly rush to stake claims on this tiny plug of basalt lava. A team of radio hams even set up a (very) temporary broadcasting station there on 16 June 2005. The little beacon has had to be renewed several times and its reliability falls far short of any navigational function, especially in winter storms. It can probably claim to be the most isolated 'lighthouse' in the world. But also the most unreliable! However it is not the responsibility of the NLB.

The Commissioners had long considered placing a beacon on Ve Skerries west of Shetland but reckoned it would cost too much. Instead they had erected one at Eshaness, 14 km (9 miles) away. On 29 March 1930 an Aberdeen trawler *Ben Doran* came to grief on Ve Skerries. The nearest lifeboat had to come all the way from Lerwick on the east coast to find the crew clinging to the mast of the wreck. Due to the appalling seas they dare not approach, so nothing could be done. After a 193 km (120 miles) voyage in huge seas the Stromness lifeboat arrived from Orkney to assist, but by then there was little left of men or boat. All nine crew perished.

A gas-lit whistling buoy was put in place at Ve Skerries in 1932 until, in 1977, the *Elinor Viking* grounded there; fortunately this time the crew were rescued by helicopter. The matter then became especially urgent when, the following year, oil tankers began using Sullom Voe, the largest oil terminal in Britain.

The challenging offshore operation used Eshaness as the supply base and all materials were taken out by helicopter. The extremely exposed rocks protrude only 2 m (6 ft) or so at high water, so holes were drilled 3–4 m (10–12 ft) down into which were fixed thick, high-tensile steel bars. These extend inside to the very top of the tower and are tensioned to exert a strong downward force holding the tower firmly in place. The NLB engineers opted for a prefabricated re-enforced concrete tower (14 m or 46ft) high on top of which was placed the lantern from the defunct North Carr light vessel. The innovative design and construction won an award for its engineers. The following year a major, automatic light was also built at Point of Fethaland, the most northernmost extremity of the Shetland mainland.

The 'outside route' to and from the Shetland Oil Terminal necessitated the construction in 1982 of further automatic lighthouses on North Rona 72 km (45 miles) northeast

Ve Skerries lighthouse, Shetland

of the Butt of Lewis (not to be confused with South Rona east of Skye) and on the tiny island of Sula Sgeir 18 km (11 miles) away. The temporary accommodation for the work force remained in place on Rona for a further winter until the smaller light on Sula Sgeir was completed during the summer of 1982. Both were commissioned on 23 March 1984, initially being monitored by UHF link by the keepers at the Butt of Lewis, but now of course from George Street in Edinburgh. They are serviced by helicopter from the light-house tender anchored in the lee of the island.

In 1824 the geologist John MacCulloch aptly described North Rona thus: 'Surely if aught on earth or rather on sea, can convey the complete feeling of solitude and desertion it is Rona.'

North Rona from the air

It has an entry in the Guinness Book of Records as once having had the most isolated community in the British Isles. Tradition maintains that the very first inhabitant was a Celtic hermit Ronan. His beautiful beehive-shaped cell is still relatively intact but a small chapel was added in medieval times. A circular turf dyke enclosed it within the burial ground. By this time a small community had established – probably no more than five families – and the ruins of their dwellings cluster together nearby. This 'village' is surrounded by an impressive array of parallel cultivation rigs and furrows known as *feannagan* or lazybeds. The island was eventually abandoned early in the 18th century after which only a shepherd remained to tend the landlord's sheep. MacCulloch met his family when he came ashore in 1819, but the last shepherd had departed by 1844. Two other men from Lewis later overwintered there but were found dead from illness and exhaustion in the spring of 1884; they are buried beside the chapel.

Just before the war the naturalist Frank Fraser Darling, with his wife and young son, spent some months on Rona observing seabirds and seals. In April 1941 a Whitley bomber made a crash-landing but the crew, and its top-secret radar equipment, were recovered almost immediately. A few weeks later the entire plane was dismantled and removed by the RAF, not dissimilar to the situation already described at the Copinsay lighthouse in Orkney.

Both Rona and the island of Sula Sgeir were designated National Nature Reserves in 1956. Two years later the Nature Conservancy (now Scottish Natural Heritage) began a

Left: North Rona lighthouse under construction, October 1982

Right: North Rona lighthouse maintenance, June 1998

study of the grey seals that assemble there to breed every autumn. The study continues to this day – by the Sea Mammal Research Unit based in St Andrews – with around 1,000 pups born each year. The island has impressive seabird colonies, and is one of four to six breeding colonies known in Britain of the rare Leach's petrel, a small nocturnal, burrow-nesting species, some of which nest in the village. It was this little colony that first attracted early ornithologists like Harvie-Brown, Barrington and Atkinson. Darling and Atkinson's accounts are well worth reading (the latter – *Island Going* – being one of my favourite books) and inspired me to make my first visit to Rona in 1971. I have been back to my favourite island now over a dozen times, undertaking a ringing programme of my favourite bird – the Leach's petrel. Indeed an adult I caught and ringed there in 1972 was recaptured back in the village in 2002 – a highly respectable age for something the size of a blackbird that lives most of its life out on the high seas!

North Rona chapel with Leach's petrels at night

Sula Sgeir has never been inhabited except maybe briefly, tradition maintains, by Brianuilte, the sister of Ronan, who built a little oratory and died there. But each year, for centuries, a party of men from Ness in Lewis spend a few weeks in stone bothies while, under licence nowadays, they harvest 2,000 young of the 9,000 gannet pairs that breed there each summer. Between 25 and 31 August 2005 they found a black-browed albatross – just as at the Bass Rock lighthouse some 40 years earlier. It would sit in the gannet colony, near the helipad, which services Sula Sgeir's small automatic lighthouse. The bird – a very rare vagrant from the South Atlantic – returned in 2006 and briefly in 2007 but has not been seen since.

Left: Sula Sgeir lighthouse with nesting gannets

Right: Gannets over Sula Sgeir

With the successful installation of beacons on Rona and Sula Sgeir, the oil industry was soon to open a fresh spotlight on the northwest of Britain. On 5 January 1993 the 90,000 ton tanker *Braer* was driven ashore in Quendale Bay, Shetland, just to the north of Sumburgh Head lighthouse. Fortunately its 35,000 ton cargo was light crude which was rapidly dispersed by some of the worst winter storms Shetland had experienced for some time, thus minimising what might have been a considerable environmental disaster. It did however highlight more deficiencies in British emergency plans for oil spills. A year later Lord Donaldson's Report *Safer Ships, Cleaner Seas* made 103 recommendations to improve marine safety, including the routing of laden tankers west of the Hebrides – the Deep Water Route – where there was more sea room to rectify situations if possible. As a result the NLB in Scotland constructed four completely new automatic lighthouses – one at Kyle of Lochalsh near the Skye Bridge and three more west of the Hebrides. Like all automatic lights

Haskeir Rocks and Boreray St Kilda from North Uist

Left: Accommodation huts on Haskeir Rock

Right: Constructing Haskeir lighthouse

they would be operated and monitored from George Street, Edinburgh by UHF/VHF radio and BT landline.

The major light on Haskeir (Hasgeir) island, 13 km (8 miles) west of North Uist – the NLB's 200th lighthouse and its most modern – employed the very first hybrid renewable energy source, with 13 solar panels and six wind generators. Minor lights were erected on Gasker (Gasgeir) just to the north, and on Shillay in the Monach Isles National Nature Reserve, also known locally as Heisgeir. The latter was placed beside the old 41 m (135 ft)brick tower, originally built by David and Thomas Stevenson in 1864 and abandoned in 1942 during the Second World War. All three units were completed in 200 days between April and October 1997, when I was able to visit them by helicopter with the NLB Engineer. On Shillay in the Monachs, Heisgeir light was mounted on a white metal-clad framework tower 5.5 m (18 ft) high, and flashed white twice every 15 seconds, with a range 16 km (10 miles). After a review

Left: New and temporary lights on Shillay, Heisgeir or the Monach Isles

Right: Reinstated light on the original Monach Isles tower

in 2005, the Board decided to increase the range of the Monach beam from 10 to 18 miles (29 km), for which the most attractive option was to move back into the old tower.

It is amusing to reflect that, immediately after the War, in 1947, a rumour had circulated that the Monach light was to be relit. A letter then appeared in the *Oban Times* from a resident in North Uist, 6 km (4 miles) to the east, welcoming the prospect that this 'beautiful light may shine out on the crofters' houses, and be a landmark for the locals' once more. This prompted an indignant reply from a lightkeeper's wife 'who suffered there for 3½ years; I sincerely hope it will never be reopened'. But, of course, with automation, there is no prospect of lightkeepers being posted there.

And so, in 2005, the NLB was proud to report:

As a result of the combination of exposed off-shore location, natural heritage value (seals and seabirds), cultural heritage value (Grade B Listed Buildings), contemporary health and safety demands and financial constraints, the project to re-light the Monach Isles lighthouse provides an excellent example of the type of work we do in the Projects Section of the Engineers Department … It is heartening to see that, despite the passage of almost 150 years and advances in navigation technology, the original Stevenson-designed Monach Isles lighthouse still has a vital role in the provision of a major aid to navigation.

With the very first hybrid solar-/wind-powered renewable energy source deployed on Haskeir, and being monitored from George Street, many more lighthouses are being converted to solar electric power, with back-up wind generators. Thus the NLB has taken on board a duty of care for the environment – for ethical, economical, legal and commercial reasons. Despite devolution, the NLB remains under the remit of the Department of Transport in London but where there had been some friction in the past, the NLB and Trinity House co-operate closely, as do both with the Commissioners of Irish Lights, with all three being active members of the International Association of Marine Aids to Navigation and Lighthouse Authorities (IALA). Although the very last Scottish foghorn was turned off at Skerryvore in 2005, the lights continue shining.

Nowadays this tiny country of ours has one of the best lit coasts anywhere. But it would not have been possible without the Stevensons. As the 'black sheep' or 'wreckling' of the family, the famous author Robert Louis once put it:

> *In the afternoon of time*
> *a strenuous family dusted from its hands*
> *the sands of granite and, beholding far*
> *along the sounding coasts its pyramids*
> *and tall memorials catch the dying sun,*
> *smiled well content …*

In her fascinating history of the NLB in *Scotland's Edge*, Evelyn Hood ends beautifully:

Future passages may be stormy … but behind them lie two hundred years of extraordinary achievement. For the safety of all they built some of the most remarkable structures ever engineered. For the safety of all was created a service demanding, and getting, self- sacrifice, courage and total devotion to duty on the part of its staff. For the safety of all, in peace and in war, the kindly Northern Lights have swept a bright path for the men of the sea.

But the days of lighthouse keepers were numbered. They had begun with a single keeper appointed to each light, until the first assistant lightkeeper was recruited in 1815. In due course a second assistant was added, bringing the complement at each light to three. Their primary duty was to keep the lamp lit during hours of darkness (under threat of instant dismissal). By day the lenses and lantern had to be cleaned, or the lamp trimmed with

Lightkeepers at the Isle of May landing, April 1968

records maintained of fuel consumption, weather and shipwrecks. Even when the lights were blacked out in wartime, the keepers remained at their post. At first Robert Stevenson had a cautionary approach to his keepers, relying on them to fall out and report on one another, or else they were liable to turn to strong drink in their isolation. But, as Alison Morrison-Low has concluded:

> The service was tightly governed by a series of rules, and lights – and keepers – were constantly inspected, often with little warning. The men who ran and serviced the lights were admirable and dedicated professionals, made redundant only by technology … The way of life of the lighthouse keeper came to an end as remote control, radio and radar all combined to make the profession redundant.

One could say that automation began as far back as 1896 when the two keepers were withdrawn from Oxcars (first commissioned only ten years before) in the Firth of Forth. The light was regulated by a clockwork timer and fuelled by bottled gas, delivered weekly from the depot at Granton. Trinity House began its programme of automation in 1910 with Gustaf Dalen's invention of the sun valve, sensitive to daylight and operating the delivery to the light of acetylene gas. New innovations also affected the design of the lamps themselves. Where once the huge array of prisms turning inexorably by powerful clockwork or on rollers over a large bath of mercury, smaller, yet more powerful lamps sufficed, many in sealed arrays of headlamps, first tested on Fidra. And even these have now come to be replaced by light-emitting diodes (LEDs) with low energy consumption, reduced maintenance, small size and long life. Unlike conventional light sources they do not become hot and are ideal for minor lights and buoys.

The real programme of automation began in 1960, with Little Ross, Rubha nan Gall and Kyleakin all de-manned. The burnt-out Chicken Rock off the Isle of Man was next, then Auskerry on the Orkney island of Stronsay, with ten others following suit that same decade, and 14 more in the 1970s (starting with 'unpopular' lights such as Dubh Artach and the Flannans). In anticipation of the re-structuring required the NLB ceased recruiting any more career lightkeepers in the mid-1970s. Where once it had some 600 on its payroll, the Service was gradually being down-sized by natural wastage and early retirement. Instead, more and more use had to be made of technical staff, auxiliaries, the two supply ships and helicopters.

The pace of automation then picked up, 1980 seeing Barra Head de-manned, quickly followed by Fair Isle North, Rattray and Sule Skerry. By the end of the decade another 25 or so had gone automatic, with Bell Rock and Bass Rock in 1988 and the Isle of May the following year. The Granton Depot near Edinburgh was closed in 1995 when the Engineering Storage and Testing Facility was moved to Oban (first established in 1904) or to Stromness in Orkney until it too closed in 2003. During the 1990s, the programme of automation came to be completed in Scotland and the Isle of Man with the final 27 or so stations. Kinnaird Head went automatic in 1991; since its very first NLB keeper, James Park, there had been 32 principal lightkeepers and James Oliver was the last. Others de-manned that same decade included Skerryvore and Pentland Skerries (1994), Muckle Flugga (1995), the Mull of Kintyre and Stroma (1996), and Hyskeir, Duncansby and Strathy Point (1997).

Amongst Hyskeir's last duty keepers was Allan Crowe. At 60 he was the oldest principal in the service and due to retire the following year. He was a fifth generation keeper, having been born in Aberdeen while his father was keeper on Skerryvore; he then spent his early life on Earraid while his father was on Dubh Artach. At 42, George Miller on Hyskeir was the youngest assistant lightkeeper in the service and, with Kenny Weir (54), would be kept on relief work for a further year before compulsory redundancy. They agreed that automation was inevitable: but while larger ships now navigated by high-tech radar and satellite, smaller vessels felt safer with a manned light. Author Christopher Nicholson reflected how:

> Thousands of yachtsmen who spend their weekends and holiday sailing around our coasts will perhaps feel that little bit less secure when they realise that the lighthouses they pass no longer contain any human observers to keep an eye on their progress – just in case the unexpected should happen.

To that must be added small, local fishing boats whose skippers would phone lightkeepers for local weather conditions. Hyskeir was one such and as an official station for the Meteorological Office, submitted reports every three hours.

The final five manned lighthouses – Rinns of Islay, Cape Wrath, North Ronaldsay, Butt of Lewis and Fair Isle South – still served as regional monitoring centres for an array of other lights. Hence they were the last to go automatic in 1998. Everything can now be controlled and monitored from George Street, Edinburgh.

The principal lightkeeper for the last four years of the Butt, Donald Michael, would end a 33-year career in the service. He had been one of the last to leave when the Flannans went automatic in 1971, but now his time had finally come. The last principal at Fair Isle South was Angus Hutchison. Although all four last stations were de-manned at the same time, it was here at the south end of Fair Isle that a quiet and dignified ceremony would take place to mark the end of an era. On the white-washed wall outside the lighthouse, a modest metal plate reads:

> Fair Isle South, the Last Manned Lighthouse in Scotland. This plaque commemorates the invaluable services of generations of lightkeepers from 1786 to 1998. Unveiled by HRH Princess Royal, Patron of the Northern Lighthouse Board. 31st March 1998.

Automation ended 211 years of NLB lighthouse-keeping in Scotland (353 years if one goes back to the original beacon on the Isle of May). But the towers and modern lights still require maintenance and monitoring. Back at 84 George Street, two small computers gave information on the status of 64 major lights, four minor lights and one buoy. If there was a problem, the watchkeeper notified the operations management staff, permanently on call,

Left: Plaque on Fair Isle South commemorating
lightkeepers at the last automation in March 1998

Right: Princess Anne unveils the plaque at Fair Isle South, March 1998 (NLB)

who could then despatch a repair team. Entering the 21st century, new software permitted over 350 Aids to Navigation – including all buoys and minor lights – to be added, with a back-up monitoring centre at the Oban depot. All buoys are now solar-powered, as are many minor lights and lighthouses. The NLB has a statutory duty to inspect harbour lights, those on oilrigs, to make sure offshore wind farms are adequately lit and all wrecks are marked. But, at the end of the day, despite modern electronic and satellite systems things might still go wrong, albeit on rare occasions. Witness the modern nuclear submarine *Astute* that went aground on a sandbank within sight of the defunct Kyleakin lighthouse in 2010! This was not a failure of the external navigational aids provided at great expense by the NLB fortunately, in this case it was more by human error.

Eddystone is probably the world's most famous lighthouse and eventually went on to become Trinity House's first to be automated in1982. North Foreland in Kent, at the southeast of the Thames Estuary, also holds a unique place. It was the first to be placed under the management of Trinity House and now was to be their very last manned station. When the final six keepers left in November 1998, they brought to an end a Trinity House tradition and lifestyle that had existed, in one form or another, for four centuries.

Trinity House calculated that the cost of automating a typical lighthouse might be some £365,000 but the savings it made would pay this back in only four or five years. Each lighthouse, or group of lights, only required periodic attention from a part-time attendant. The focus of attention now moved to the Operational Control Centre in Harwich, Essex. In their impressive volume *The Lighthouses of Trinity House* (2002) Richard Woodman and Jane Wilson commented that:

The Coastwise Lights of England and Wales continue to be 'kept' with the same level of assiduity as before, but the monitoring personnel involved are detached from the stations in their care. Their lives are not intimately bound up with their distant charges, nor by the wind and sea that lash them. They are tied to different routines and much may have thereby been improved, but much else has been lost in this unavoidable march of progress.

Automation meant the loss of human eyes and a reaction in the event of shipwreck, oil spillage, illicit entry or drug smuggling, but the NLB always maintained that a lighthouse is not, and never has been, a rescue centre or co-ordination point for pollution control. The keepers were always willing to help where possible, but always conscious that their primary duty was to maintain the best possible aid to navigation, thus preventing shipwreck or pollution rather than tidying up after it. Christopher Nicholson feared that a lighthouse without its keepers becomes 'simply an impassive stone monument … lighthouses themselves become just maritime curiosities to future generations.'

We must hope that this will never prove true. Woodman and Wilson are more optimistic in outlook:

> While the mariner enjoys the assistance of electronic aids on the bridge of his ship, and while these include the impressive and highly accurate Global Positioning System (GPS), a failure of any of these can prove disastrous. Should such an even occur, either from external causes such as the destruction of the GPS satellite constellation, or an internal cause such as a power failure, the lighthouse remains as a fail-safe and reliable seamark. Few other 'inventions' have such a long history, for the lighthouse is almost as old as that of seagoing ships themselves.

Certainly that same loss of human presence at lighthouses had consequences for natural history. The innovative and comprehensive migration surveys utilising lightkeepers initiated by Harvie-Brown, Cordeaux, and Barrington, would never again be possible, for example. Curiously botany seems never to have been a popular pursuit amongst keepers – other than gardening of course. The observation of animal behaviour, the reporting of unusual or rare species, of weather and dramatic sea conditions, even earth tremors have all but ceased, especially in the bleak winter months and from the remotest of stations.

But all is not entirely lost in such respects and happily, in many cases, naturalists have moved in – albeit seasonally perhaps – to redundant lighthouse buildings, and in a few cases, to the lighthouses themselves. We have already mentioned the establishment of the Isle of May Bird Observatory in 1934, to occupy the disused Low Light; the Isle of May is a National Nature Reserve and open to visitors. There are also observatories at Copeland, Bardsey, Lundy, Portland and other lighthouses. One of the old Nene lights in Lincolnshire was for some years the home of the great naturalist, conservationist and painter Peter – later Sir Peter – Scott, the founder of the Wildfowl and Wetlands Trust. In 1990 Aberdeen University established a Field Station in the old Cromarty lighthouse for the study of dolphins and other marine mammals. The island of Copinsay was bought in 1973 as a

The original lighthouse on Lundy, Bristol Channel, built in 1819

memorial to the noted ornithologist James Fisher and is operated as a bird reserve by the RSPB; appropriately its sandstone cliffs, directly below the operating lighthouse, support the second largest seabird colony in Orkney. The old Muckle Flugga shore station at Burrafirth on Unst, in Shetland, is now a visitor centre for the Hermaness National Nature Reserve while a similar £5.4 million project at Sumburgh Head, at the opposite end of Shetland, will doubtless prove an exciting tourist attraction: this is a superb location to view the cliff-nesting seabirds, seals and cetaceans such as orca. There are wonderful wildlife viewing opportunities provided by the RSPB at Dunnet Head lighthouse in Caithness, at Rathlin's 'upside down lighthouse' off County Antrim and at South Stack in Anglesey, amongst others. These boast some spectacular seabird cliffs while the South Stack usually has

Public viewing of the impressive seabird cliffs at Rathlin lighthouse, Co Antrim

CCTV cameras in operation – at nesting razorbills and rare choughs for instance. At some locations too, (including South Stack, Kinnaird, Ardnamurchan, North Ronaldsay, Kintyre and Kyleakin) there is access to the lighthouse itself. Many visitor centres are manned by wardens in the summer, who routinely record wildlife sightings. Other stations have been converted to museums (such as Kinnaird), religious retreats (Holy Island for example), hotels (Corsewall), remote restaurants (such as Eilean Glas, Scalpay – albeit short-lived) or else offer holiday accommodation (Fair isle South and Sumburgh Head), even art or field study courses (Rubh Re for example). I recall one particular fantasist, totally ignorant of the unrealistic logistics at such a remote site, considered turning the accommodation block at Barra Head into a recording studio! The list will continue to expand and diversify as more redundant lighthouse buildings are sold off. Thus, although pairs of human eyes have been lost at the most isolated stations, the role of observant lightkeepers at others has been taken over by naturalists or visitors themselves.

Mariners still find the blinking beam of a lighthouse useful – and indeed reassuring. At the end of the day, who could deny that in the age of the silicon chip we have yet to find an adequate replacement for a lighthouse? The microprocessor however now seems to be an adequate replacement for the lightkeeper himself and only requires maintenance twice a year, instead of twice a day with a lantern. As Bella Bathurst somewhat discourteously quipped: 'at the tail end of the twentieth century it does not require three grown men to keep a light bulb.'

But as an Orkney ferry captain once put it: 'This GPS tells me where it thinks I am, that lighthouse or that buoy tells me where I know I am, and I am much happier. In those circumstances when weather and visibility is bad then I have that confidence of knowing where I am from those traditional aids.'

Another ocean-going ship's master has added: 'I have all the modern navigational equipment on board, but there is nothing more friendly and welcoming than the sight of a flashing lighthouse.'

The reassuring flash of Hyskeir lighthouse from the Oban to Uist ferry

POSTSCRIPT

In 1989 the Isle of May – Scotland's oldest light station – became automatic and, for the first time in 353 years, no keeper was resident on the island. It became uninhabited except for the Nature Reserve seasonal wardens, researchers and the periodic presence at the observatory. There has been a long-term study of puffins on the island while scientists have also looked at other seabirds, mice, seals, insects and so on, while of course migrants and vagrants continue to be added to the bird list. But in 2006, and again in 2009, the reserve wardens paid host to a very special human visitor. We may recall the tragic deaths from suffocation of the lightkeeper George Anderson and his family on the last Sunday of January 1791 – father, mother and four children. An infant daughter, Lucy, not yet a year old, was found alive three days later (see end of Chapter 4).

That very special visitor was Susan Ciccatelli, a great, great, great, great granddaughter of Lucy Anderson. Scottish Natural Heritage's reserve manager exclaimed: 'We are used to having unusual bird visitors turning up – but unusual human visitors are something else!' Susan Ciccatelli told the reserve wardens: 'Eventually I inherited a lot of family papers and was struck by the ones relating to Lucy. I gathered all I could from the internet and started to piece together the details of Lucy and her family'. The National Library and the internet yielded a lot of information, including the account of the tragedy in the Edinburgh *Evening Courant* newspaper, but there was no mention anywhere of where the family was buried. Despite exploring old graveyards in Fife, Susan has been unable to find out where George Anderson, his wife and his four children are buried. She thinks that the family might be buried on the Isle of May but there is no record of this.

A remarkable story indeed. Yet, far away in America the inscription on a grave monument in the cemetery at Andes, New York, simply reads, 'Our mother Lucy Anderson, wife of Henry Dowie, born on the Isle of May, Firth of Forth, Scotland.'

Bishop Rock, Scilly in front of the setting sun

A Saga of Sea Eagles

...This is an important little book, as much for its wealth of information and insight as for its value as a very detailed historical record. It is superbly illustrated, with masses of colour photographs and lots of the author's wonderful drawings. ...we should be enormously grateful to John Love for pulling all these strands together for us, and for doing it so well. *IBIS*

The re-introduction of sea eagles is a great conservation success story, and John Love has played a vital role in it since the 1960s. He is an expert on the subject, but has chosen to write a memoir rather than a scientific report, and the result is superb. His book is full of anecdotes, humour, vivid personalities and dramatic places - St. Kilda, Rum and Fair Isle all feature. *Scotland Outdoors*

978-184995-080-0   £19.99

NPC — 26/3/21